Digital Exhaustion

Digital Exhaustion

SIMPLE RULES FOR RECLAIMING YOUR LIFE

PAUL LEONARDI

RIVERHEAD BOOKS

NEW YORK

2025

Riverhead Books
An imprint of Penguin Random House LLC
1745 Broadway, New York, NY 10019
penguinrandomhouse.com

Grateful acknowledgment is made for permission to adapt and reprint portions of "Helping Employees Succeed with GenAI" by Paul Leonardi that first published in *Harvard Business Review* (November 2023). Used with permission of Harvard Business Publishing.

Cartoon by Liam Walsh on page 156 and cartoon by Amy Hwang on page 239 used with permission of www.CartoonStock.com. Cartoon by Marketoonist on page 251 used with permission.

Book design by Daniel Lagin

LIBRARY OF CONGRESS CONTROL NUMBER: 2025020803

ISBN 9780593851234 (hardcover)
ISBN 9780593851241 (ebook)
ISBN 9798217179091 (international edition)

Printed in the United States of America
1st Printing

The authorized representative in the EU for product safety and compliance is Penguin Random House Ireland, Morrison Chambers, 32 Nassau Street, Dublin D02 YH68, Ireland, https://eu-contact.penguin.ie.

For Norah, who is almost never exhausting

CONTENTS

PART I
The Exhaustion Triad

PART II
Simple Rules for Resilience

PART III
Complex Contexts

CONTENTS

PART III

Complex Contexts

Sigh, Scroll, Click. Repeat.

In June 2000, I stepped off the escalator at the Montgomery Street BART station in San Francisco's financial district for my first office job. Looking up at the skyscrapers lining Market Street and seeing the rush of people all around was exhilarating. That's one of two memories I have of that day. The second memory is of my new boss's face. His name was Brian. He beamed when he greeted me with a warm handshake and a hearty "Hello." I don't recall what we talked about—probably how to log in to my computer or where the coffee machine was—but I haven't been able to forget the energy and enthusiasm he radiated as he told how much he loved his job and how great the company was.

I don't think I would remember that image so well had it not contrasted so sharply with his disposition at the end of the day. When I walked into his office to say goodbye, I had to do a double take to make sure he was the same guy I'd met that morning. He was slumped in his chair, eyes glazed over, staring vacantly at his screen. He was scrolling, endlessly it seemed, through pages of text, and I can still hear the sigh he made before he dropped his head. Then he noticed I was standing there. He tried to perk up and appear energetic, but it didn't work. He was clearly exhausted. Just about every day played out the same: I met perky Brian in the morning and bid farewell at the end of the day to a defeated man.

A few weeks into the job, I got up the courage to ask him about his daily transformation. "Is it that obvious?" he asked, clearly a bit embarrassed. "I love my job, don't get me wrong," he said. "But all day long—the email, the calls, the system data, the reports—it just wears me out. I get to a point in the day where I'm just staring at the screen feeling like crap because I know in the morning I was super focused and productive, but by the end of the day I'm beat." And then he said something that has been ringing in my ears for more than two decades: "It's like I'm depleted. I'm not physically drained because I can go straight to playing basketball after work, no problem. And it's not like I'm burned out of the job. But something about all the technology and all the inputs coming at me all day just makes me exhausted."

I understood how he felt. Even though it was my first job, I felt it too. So many technologies to learn. So many streams of information. Eyes glazing over as you stare at the screen. The aimless scrolling. Knowing what task you should be doing but not being able to start it. The feeling that no matter how many messages you respond to, how many posts you read, or how much data you sort through, you can never keep up. And perhaps worst of all, that depressing realization that this morning you were more focused and energetic than you are now and that yesterday you seemed more on top of it than today.

Two years later, I was in graduate school. By that point I'd interviewed nearly one hundred people who worked in what we would typically call knowledge jobs across a range of industries, from banking to education to consulting to marketing. No matter who I talked to, these people's stories about their work eventually started to sound like

Brian's: They liked their jobs and they were excited about their work, but they ended their days feeling a sort of exhaustion that was hard to describe. The one common thread I could discern was that their description of exhaustion seemed most intense when they talked about the digital technologies they used at work and at home.

So at the start of 2002, I decided that in every study I did, I would ask a simple question: "How much do the digital technologies you use make you feel worn out?" Sometimes I would sneak this question in at the end of a survey, and other times I'd ask it casually in an interview. I always had people rank their response from 0 (not at all) to 6 (so worn out that I can't even keep looking at the screen). I wanted to know who felt that their digital tools were exhausting them and understand why they felt that way. Over the years, no one has ever had a problem answering this question, and rarely has someone asked me what I mean or to define "worn out." The question clearly resonated.

Fast forward to COVID-19. At the height of the pandemic, when most countries had shut down and nearly all knowledge-intensive jobs had gone remote, the daily news began to explode with stories of people's strained romance with their technologies. A study exploring "Zoom fatigue" got lots of press, *The New York Times* published a well-read article called "It's Time for a Digital Detox (You Know You Need It!)," and the World Economic Forum titled one its reports "Are You Suffering from Digital Exhaustion?" All my friends and coworkers ratcheted up their complaints about the amount of time they spent on digital tools, and the people I interviewed and surveyed seemed to bring up the topic of how crappy their technologies made them feel before I could ask my usual question about it.

By this time, I had spent nearly twenty years learning about what caused digital exhaustion and how people effectively coped with it, but I had never tabulated the responses. It was time to look at the data. So I did.

Ratings of Digital Exhaustion from 2002 to 2022

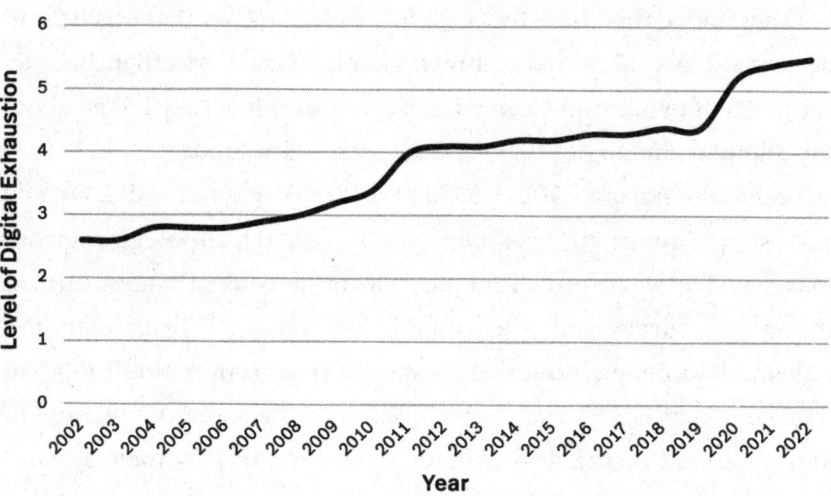

This graph shows the average responses of 12,643 adults from twelve countries spanning the twenty-year period from 2002 to 2022. It includes people aged twenty-one to seventy-five who worked in more than fifteen different industries and in all types of roles. The years I collected data are displayed along the horizontal axis. The vertical axis presents the scores (from 0 to 6) that people chose to represent their level of digital exhaustion. In 2002, the average exhaustion score reported by 426 people was 2.6, just below the midline of feeling worn out by digital technologies. But by 2022, the average exhaustion score reported from 739 people was 5.5, indicating that respondents were feeling extremely worn out. When I first showed this figure to my family, I thought they might comment on how clear the trend in the data was or tell me they were impressed that I'd asked the same question of so many people for so long. Instead, my then-nine-year-old daughter said, "It looks like a snake about to strike."

The data reveal a disturbing trend: People have been growing increasingly worn out by their use of digital technologies at work and at home. Importantly, this trend did not start with COVID-19, nor did it end when lockdowns were rescinded and we began to circulate in the world again. Even in the early 2000s, before the advent of social media such as Facebook, Twitter, and YouTube, people were experiencing moderate levels of exhaustion from high digital tool use. Digital exhaustion has been with us for a long time. But you can see two important jumps in the graph.

The first jump occurs between 2010 and 2011. This was a period of major transformation in our digital landscape. Facebook and YouTube's monthly active users (a key metric tracked for online subscription-based products) jumped to more than five hundred million, nearly double the number of users two years prior. During this time, smartphone users reached more than one hundred million in the US alone, giving a historically large number of people constant connectivity to online content and social networks. Our use of digital technologies to access data and connect with others had been growing for some time, but the twin forces of software companies making their money by selling our eyeballs to advertisers and device manufacturers getting into our pockets and purses created a leap in our feelings of exhaustion.

The second jump is, not surprisingly, timed with the spread of SARS-CoV-2 and the associated worldwide shift to remote work in 2020. Working from home; talking with our coworkers, friends, and family on Zoom, FaceTime, or Microsoft Teams; and the blurry boundary—if not the total destruction of the boundary—between work and home wore us out. What is more surprising than the increase, though, is that although the world reopened and there is a general societal awareness that the tools that keep our digital economy humming are depleting us, the last few data points show no drop in exhaustion levels. To the contrary, it seems like the snake of digital exhaustion isn't waiting to strike. It has struck. And with a vengeance.

In today's world, if we want to work an office job, be a good friend or sibling, interact with most of our civic institutions, and maintain relationships with people near and far, we can't escape digital technologies and the threat of exhaustion they bring. But we can learn to use them in healthier ways.

In an attempt to combat this problem, it's become popular to talk about adopting a philosophy of "digital minimalism," as author Cal Newport calls it. Much like what I've heard from so many digital technology users, Newport observes that "almost everyone" he spoke to in his research "felt as though their current relationship with technology was unsustainable—to the point that if something didn't change soon, they'd break too. A common term I heard in these conversations about modern digital life was *exhaustion*." I'm completely on board with the idea of digital minimalism. The more we can minimize our use of the technologies that create the conditions for our digital exhaustion—while still reaping their benefits—the better off we are likely to be. Some suggest that a good strategy is to embark on a "digital detox," permanently eliminating—or if you can't do that, taking an extended vacation away from—your digital devices. Although a hiatus in technology use has some benefits, neither I nor the vast majority of people I've interviewed and worked with over the last twenty years have found this to be a realistic long-term option.

Even if you could take time for a digital detox, the problem is that all vacations must come to an end. "Leaving Las Vegas for ten days if you're a problem gambler is great," says Alex Pang, who has written books on distraction and the importance of rest. "But if on day eleven you're back at the slot machines, then it's not so great." The evidence shows that a vacation from your phone or your computer probably won't solve much. In my own research, I've found that people who take a significant amount of time away from their digital technologies have a bumpy reentry once they decide to return to them. The world kept turning while they were gone and there is now too much to catch up on. The strain of working extra hard to recoup lost time, coupled with the blissful memory of a vacation away from these devices, can lead to even more desperate feelings of exhaustion.

There have been many reports about our direct physiological responses to our digital devices that discuss how staring at screens can cause our eyes to fatigue or how the blue wavelengths emitted from them can wreak havoc on our circadian rhythms. These physical effects, though important, are not the subject of this book. Instead, my goal is to understand how the access to information, data, and people that our digital devices enable contributes to our digital exhaustion. Devices are not the problem; the way we use them and the social, organizational, and cultural expectations associated with our patterns of use are.

In today's digitally connected world, our secret weapon is knowing how to develop a healthier relationship with our technology. We can't stop using our devices, so we have to learn to use them in ways that don't sap our energy and, importantly, in ways that give us new energy. By taking back control of our digital tools, we can use them for the very reasons we adopted them in the first place: to help us connect better with others and to be more creative, more efficient, better at our jobs, and happier people. Those are the promises of digital technologies. Reimagining our relationship with them so that they stop exhausting us is how we fulfill those promises.

Let's begin by discussing what digital exhaustion is.

EXHAUSTION UNDER THE MICROSCOPE

The feeling of digital exhaustion is easy to recognize but much harder to define. Anna Schaffner, a professor at the University of Kent, sought-after exhaustion coach, and the author of the book *Exhaustion: A History*, reminds us that although Western societies have written about the concept of exhaustion since at least 350 BCE, it has never been well defined. In surveying more than two millennia of writing on the topic, she concludes: "Exhaustion generally suggests the vampiric depletion or harmful consumption of a limited (and usually nonrenewable) resource, which leaves an originally well-functioning person, object, system, or terrain in a weakened or dysfunctional state." I love the term "vampiric." Even though exhaustion may be hard to define precisely, the image of an insidious beast that sucks the energy from its host seems on the mark.

There appears to be general agreement that the symptoms of exhaustion include weariness, disillusionment, apathy, hopelessness, restlessness, irritability, and a general lack of motivation. Exhaustion is simultaneously a mental and physical phenomenon. When we overtire the brain, our body attempts to reserve its energy to restore our cognition, and we feel physically drained. Exhaustion reduces our energy, our desire to act, and our ability to focus and concentrate. Mental exhaustion is a particularly acute problem because unlike our bodies, our brains rarely send clear signals that we are tired. It is not always easy to tell when we have hit our limit and need a break. If you're doing something physical like moving boxes or swinging a hammer, your body will tell you when it is tired and needs rest. We are good at interpreting the signals of fatigue coming from our muscles, recognizing pain from our joints, and noticing that we are not hitting nails with the same force or accuracy that we were earlier in the day. But we have much greater difficulty recognizing our mental exhaustion. In the moment, we might be aware that we are making more mistakes than usual. But usually, it's the physical symptoms like a stiff neck, a sore back, or dry eyes that tip us off to the fact that we are mentally exhausted. We often push our brains harder and longer than we can push our bodies, and, in a cruel paradox, the more we overwork our brains and deprive them of rest, the less likely they will be able to rest when we want them to.

Now let's add digital technologies into the mix. We live in an age of information abundance. No matter where we go or where we look, we are overwhelmed with data. At work, email, Excel, databases, text messages, Slack, Zoom, ChatGPT, and so many other digital technologies that permeate our everyday lives inundate us with information and data. A growing number of research studies on the effects of these so-called productivity tools show that our increasing use of them leads to exhaustion. Away from work (or maybe secretly at work!), the sites on which we do our shopping, like Amazon and Alibaba, music platforms like Spotify and Pandora, and video streaming services like Netflix, Hulu, and Disney+ provide us with endless opportunities to consume. The research shows that these sites exhaust us too because they provide so many options and so many reviews to wade through that choosing is overwhelming.

No type of technology in the digital environment has been so closely linked to exhaustion as social media. The messages. The photos. The videos. The memes. The likes. The recommendations. The news alerts. The windows into other people's lives. The advertisements. Participating in social media is like drinking from a data fire hose. But it's not just the quantity of data that exhausts us; it's the fact that we can't escape. What is clear by now is that social media platforms like TikTok, YouTube, Facebook, Instagram, Snapchat, Reddit, and others are extremely addictive and designed to be so. That urge to return to our devices and be exposed to more and more data on more and more occasions is what can so easily lead us to exhaustion.

In our quest to define digital exhaustion, it's useful to pause for a moment to situate exhaustion in relation to two other well-known concepts: stress and burnout.

Let's start by drawing a distinction between stressors and stress. Stressors are external events or conditions that can trigger a stress response in our bodies. They are things that we typically perceive as a threat or challenge to our physical or emotional well-being. Some common examples of stressors include work-related pressure, financial difficulties, relationship problems, health issues, or major life changes such as a move, a new job, or the birth of a child. When you experience a stressor, your body undergoes a complex biological and physiological response known as a "stress response." Prolonged exposure to stressors can lead your body into the exhaustion stage of stress response. In this stage, the body's resources are depleted, and you experience fatigue and decreased ability to cope with stress. If we experience too much stress over too long a period, no matter whether it's good or bad, our minds and bodies stop working at peak efficiency because we are worn down.

While stress can be a precursor to exhaustion, burnout is one of exhaustion's most fiendish consequences. Burnout is normally defined as a state of emotional, physical, and mental exhaustion caused by prolonged or excessive stress. Most often, burnout is associated with our jobs and the stressors that kick off the stress-exhaustion-burnout chain are thought to be found in the workplace. The most current research suggests that burnout is recognizable through three characteristics: exhaustion, cynicism (sometimes called "depersonalization"), and inefficacy. Although exhaustion is a key contributor to burnout, burnout is bigger and more complex than exhaustion. But that doesn't mean exhaustion is unimportant. In fact, research by the psychologist Christina Maslach has shown that exhaustion is the key driver of burnout because it sets people on a trajectory to depersonalization and reduces feelings of self-efficacy. She and her colleagues write, "Exhaustion is not something that is simply experienced— rather, it prompts actions to distance oneself emotionally and cognitively from one's work, presumably as a way to cope with the work overload."

Fixing exhaustion alone won't fix burnout because people's job assignments, their workload, and their relationships with their colleagues and boss, along with contextual factors like whether they work remotely or in the office, all contribute to their feelings of burnout. But it's also true that to deal with burnout, we must get a handle on exhaustion.

Exhaustion can come from many places. It's my aim to convince you that a significant portion of your exhaustion comes from the way you use your digital technologies and that you have to reduce the stressors that cause exhaustion before it becomes a contributing factor in your own burnout.

LEVELS OF DIGITAL EXHAUSTION

A few years ago, I started a consulting job with a major software development company that built digital collaboration tools for medium and large businesses. Throughout the project, I met on the first Tuesday of each month with a product manager named Andi. She was in her early thirties and full of energy and enthusiasm for the project. In the course of our conversations, I learned that she had recently joined the company after being recruited from a small start-up that had built a competitor product. Andi was the ideal product manager. She had great technical depth and understood how to move a project forward. In every meeting I saw her lead, she was attentive to her team members and worked hard to make their jobs interesting and fulfilling, while also staying on time and on budget.

But just over a year in, I'd seen a dramatic change in Andi's disposition. She seemed to lack the excitement for the product that she once had. When I saw her interacting with her team, she seemed short on patience and somewhat irritable. One morning I asked if she was doing OK. "Not really," she responded. "I just can't keep up with it all. I've never had this experience before. I get so exhausted from going from this task to that task, to this tool to that tool, to dealing with this data and that data. It's too much. I find myself staring at my screen a lot, scrolling and clicking and generally unmotivated. I feel like my battery is worn out and no matter what I try, it can't hold a charge anymore." Not being one to miss an opportunity for data collection, I asked her, "So, on a scale of zero to six, how much do the digital technologies you use make you feel worn out?" She laughed for a second and then said, "Six, for sure. But why doesn't the scale go higher?"

Over the years, a number of people have described their exhaustion to me using the language of a depleted battery in need of a recharge. It's a metaphor that I too have come to find useful in talking about digital exhaustion—and in dealing with my own. We have lots of experience with batteries because they power the devices we depend on, like our phones, laptops, and, increasingly, our cars. A battery has only so much capacity. It can hold only so much energy, and the rate at which that energy is exhausted depends on the demands we make of the device it powers. If we use our phone a lot, or if we use many power-hungry applications on our phone simultaneously, we'll drain the battery faster than if we use it more sparingly. When we start our day well rested and enthusiastic, we feel charged. But with each email we receive, each report we read, each friend's picture we comment on, each choice we make about what playlist to select, we use up that valuable energy we have stored. The more we engage with the data, information, and people that our digital devices connect us to, the more we feel depleted. We know when we are exhausted because we recognize the signs we discussed earlier. Sometimes we can go weeks before we feel exhausted, sometimes it takes just a few hours.

This normal process of going from full charge to energy depletion is what I call "Level 1 exhaustion." We expect our energy reserves to fall based on our activities, and we know we need to recharge to restore them. Our use of digital technologies is, of course, not the only thing demanding our energy. But as I will demonstrate in part I of this book, our use of them is becoming the central driver of exhaustion today because so much of our lives is mediated by our technologies, and our choices about how we use them have a direct impact on how much of our energy they demand. Choice is important. In each recharge cycle, we can choose how we expend our energy. Just like we can choose not to run our Wi-Fi and flashlight, listen to music, and scroll through Instagram at the same time, we can also choose to engage with our digital devices, and the content and people to which they connect us, in ways that demand less energy. We have a great deal of agency in choosing how long our energy will last before we need to recharge.

The reason these choices are so consequential is because the number of times we can expend energy and recharge is limited. The constant cycle of depletion and recharge that characterizes Level 1 exhaustion takes its toll. Over time, we notice that our energy doesn't last as long as it used to and find ourselves more anxious, irritable, and demotivated. This chronic experience is what I call "Level 2 exhaustion."

We see signs of Level 2 exhaustion all around us. A 2023 survey of 3,400 people across ten countries showed that 43 percent of all employees reported that they felt "often" or "always" exhausted, and 46 percent of middle managers—those in the workplace most affected by digital exhaustion—said they planned to quit their jobs within the next twelve months because of work-related stressors. Jivan, a technical sales manager I talked with, is a member of this group. He is in his late forties, recently changed jobs, and started working with a therapist to deal with his Level 2 exhaustion. He explained it to me like this: "I got to a breaking point where I couldn't even do simple tasks right. I was too worn out from being worn out all the time. Feeling exhausted by all the information and the contacts with people and the switching between this and that and then needing to find energy to do it all again was hard. But finding the energy to come back again and again got to be too much."

A battery's health is typically measured in terms of its cycle life, which refers to the number of times it can be charged and discharged (Level 1 exhaustion) before it starts to lose its performance (Level 2 exhaustion). Lithium-ion batteries (like those that power our phones and laptops) typically have a life of around three hundred to five hundred cycles before they start to degrade significantly and eventually die. The good news is that although the battery metaphor helps us to understand exhaustion, we are not consumer batteries. If we experience Level 2 exhaustion, we can bounce back. But the metaphor is still useful. Our job is to reduce the number of Level 1 exhaustion-recharge cycles. To do that, we have to learn how to conserve our energy and find new sources of energy so that we don't get so depleted in the first place. To do these two things, we must reimagine our relationship with our devices.

SOME SIMPLE RULES

If there were one universal law about the use of digital technologies, it would be this: People settle into patterns of use quickly. Whether that fact delights or terrifies you, the evidence suggests that most people's experimentation with new technologies is short-lived. After about twelve to sixteen weeks of testing the capabilities of new digital tools, most people will have settled into a routine with them, and the ongoing effects of their use will be fairly predictable. The duration of that window of uncertainty (or opportunity) has been constant for the past thirty years. The arrival of social media and AI-based technologies using large language models (LLMs) has not only prevented the window from shutting when it should but has shattered the glass. Unlike virtually any other digital technology we're accustomed to using, AI-enabled digital tools (and just about every digital tool being designed and sold today is embedded with AI capabilities) are designed to change by themselves—continuously. Each time you provide new data to an LLM to produce text or computer code for you, the technology learns and its capabilities grow. That means that you are never really using the same technology twice. The things it can do for you this week will change by next week. Thanks to the autonomous learning that characterizes even today's most basic digital tools, you aren't learning to use a new technology once—you're learning to use a new technology nearly every time you engage with it. In short, we live in an increasingly fast-moving and unpredictable digital world.

Kathy Eisenhardt, a professor at Stanford's School of Engineering, has spent nearly four decades studying how companies succeed in chaotic high-velocity markets and is widely regarded as the foremost authority on technology strategy. Her work shows that companies that do best in environments characterized by rapid technological change have one thing in common: They rely on a set of simple rules. When markets won't stand still because technologies are constantly evolving, trying to develop a detailed strategy is futile—by the time you have your strategy ready, the market is no longer the same and all your hard work is for naught. But when companies develop, follow, and stick to a set of simple rules, they know what to do when change happens and they're prepared to adapt.

This strategy is a useful guide for thinking about how to reimagine our relationship with digital tools. If the applications and devices we use are changing rapidly and we don't know what capabilities and liabilities they'll present to us, we will be in the strongest position to avoid digital exhaustion if we have some simple rules to follow.

To determine what simple rules are likely to be most effective, we first need to get a handle on where our digital exhaustion comes from. In part I of this book, I discuss the roots of the problem, what I call the "exhaustion triad": the three major forces that shape our digital exhaustion. These chapters examine the new energy sinks of our day, showing that the ways we pay attention, how we make inferences of ourselves and others, and the emotions we feel when we're in front of our screens work interactively to sap our energy. This is an important place to begin because we can't moderate our energy expenditure or find the right ways to recharge if we don't understand what's draining us. By the end of part I, you'll see digital exhaustion in a much more nuanced way than simply as something that comes directly from our devices.

Part II presents eight simple rules for digital resilience you can follow to reshape your relationship with the ever-changing cadre of technologies you use at work and home. Just like the feature on your phone that limits apps from running in the background, changes in how you use your digital technologies can help prevent Level 1 exhaustion. The chapters in this section discuss how to make these changes and build routines around them that will help them last. Some of the rules presented in this section will help you to reclaim your energy and enthusiasm by being deliberate about your digital technology use. I'll show examples of new uses for digital tools that will reenergize you, and I'll walk through some of the latest research about how you can find consistent hits of non-digital dopamine that will encourage you to use your devices only for the things they're best equipped to assist with.

In part III, we will explore how to leverage these rules in three different contexts: at work, while parenting, and when interacting with AI. We'll look at how we can create a healthy culture of technology use in our workplaces, and we'll examine the causes, effects, and solutions to the problem of digital exhaustion that are increasingly common in hybrid and remote work arrangements. We will also examine how parents can use the rules from part II to reduce their own digital exhaustion as they attempt to navigate a world in which they are expected to do more and be more than ever, while also setting a good example for their children about how to use digital technologies in healthy ways. Finally, we'll explore the emerging role of AI and what effects it's likely to have—good and bad—on our exhaustion. The sudden introduction of AI into our lives will have major implications for our exhaustion not only because AI represents yet another set of digital tools with which we can interact but also because the way AI deals with data and presents it to us is unprecedented. We'll also explore how AI can be used to minimize our digital exhaustion and how to make sure it doesn't become yet another energy consumer.

It's all too easy to vilify our digital tools and want to separate from them when we start to recognize they are major contributors to our overall exhaustion. But it's important to remember that we started using these tools for a reason. In most cases, they allowed us to do something better or something new. That's exactly what new technologies should do for us. And, of course, many of them do allow us to improve some aspect of our lives. But that constant access to information, to other people, and to the manifold distractions that devices bring exhausts us. Again, it's not the technologies that are the problem, it's how we use them. And to make things more difficult, we don't always get to choose how we use them. Our bosses, our coworkers, our friends, our kids' friends' parents, and so many other people lure us into relationships with our digital tools that can wear us out.

The good news is that despite these constraints there is a simple path to reducing our digital exhaustion. Of course, just because it's simple doesn't mean it's easy. But by understanding why technology drains us, following the evidence-backed rules outlined in this book, and knowing how to apply them in different contexts, you can be well on your way to defeating digital exhaustion and leading an energized life. It's time to reimagine our relationship with our devices.

Digital Exhaustion

PART I

::::::::::::::::

The Exhaustion Triad

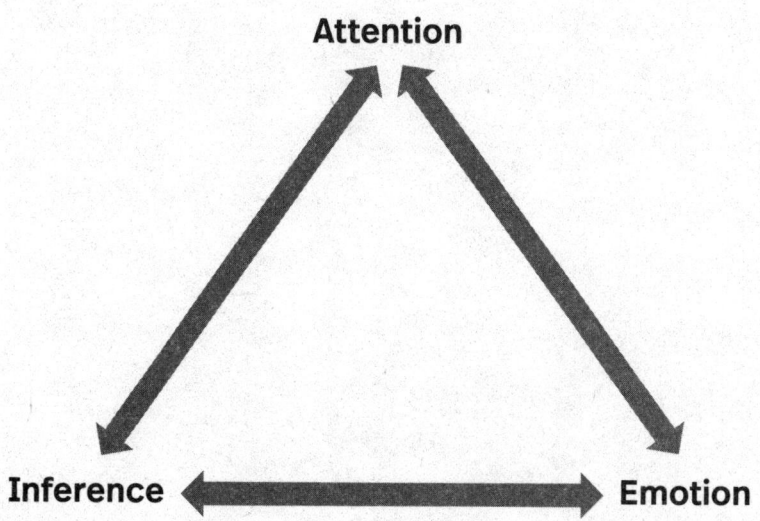

CHAPTER 1

Attention:
You Really Do Pay with It

Maya is jolted awake at 5:50 a.m. by the buzzing in her right ear. The buzz and the light accompanying it emanate from her smartphone, which sits on her nightstand, inches from her resting head. She grabs her phone, turns off the alarm, and looks at the time. As on most mornings, she groans and flops the phone back down on her nightstand. But seconds later, even though she is still half asleep, she instinctively reaches for it, enters her passcode, and taps the icon for the Instagram app. She spends several minutes scrolling through pictures posted by her friends and family members— as well as strangers she's never met—before a notification pops up alerting her that she's received an email from *The New York Times* previewing the day's top news stories. She clicks on the notification, which opens her email application. Before she can even look at the *Times*, she notices several emails that arrived from colleagues in Europe while she was sleeping. She clicks on the first one and begins to read it while her partner, who is lying on the other side of the bed, groans something she doesn't hear. She's about to ask him what he said, when the "do not disturb" mode she set on her phone turns off and she sees she has three text messages and one WhatsApp message. She clicks on one of the texts, but before she reads it, she thinks perhaps the WhatsApp message might be more important, so she switches to that app. It turns out it's just a group message from some friends overseas. So she switches back to her texts and reads one from a close friend asking her if she can believe the recent Supreme Court ruling. Her partner grumbles something again, and Maya still doesn't hear what he wants because she's too busy clicking on the *New York Times* app to see what news she can find about the Supreme Court. "Shit," she says as she notices the time (Maya tells me she makes this exclamation just about every morning). She hops out of bed, turns on the shower, and just as she's ready to step under the warm water she finally hears what her partner is asking on his third attempt to get her attention: "Hey, Maya, do you know where my phone is?"

Maya's story could be my story, your story, or anyone's, really. There's hardly a moment in the day that's free from some news source, application, or digital device trying to catch our attention. We live in an attention economy. Our attention is valuable because it's a limited resource that companies can monetize. Maya knows that if she clicks on one of the ads for Skims exercise clothing that seem to have overtaken her Instagram feed, Kim Kardashian's company will pay Instagram a small fee for the referral. But what she and most other people don't fully appreciate is that there is an entire set of lucrative economic transactions that don't depend on whether we click on an ad; the owners of the website or application are getting paid if they can provide proof that we simply saw the ad. This is what the world of digital marketing refers to as "impressions." If Instagram or *The New York Times* or any other commercial website can show advertisers that people like you and me are likely to stay on their site for a long time, they can charge more for the ad placement because more time on the site means a higher probability of the ad making an impression on you. From the moment you wake up, the attention merchants are angling for you.

But it's not all about money. Those emails from work colleagues in Europe and the texts from friends are also vying for your attention. So is that voice coming from the other side of the bed. We receive and send so many texts and emails (and have to call from the other side of the bed multiple times) because we know that other people's attention is limited and we are fighting to get a piece of it. And the data show that inundating people works. One of the first research studies I conducted was with project managers from six different companies. Project managers often have the unenviable task of trying to coordinate people and get them to do things without having direct reporting authority over them. That means they are in a constant battle to grab and hold people's attention. We found that project managers who were the most successful in moving their projects forward on time and on budget were those who sent the same message to people on their team multiple times through different technologies. They would email, call, and walk by colleagues' desks to tell them the same thing three times through three distinct media. And it worked. Project managers who engaged in this kind of redundant communication were able to cut through the various demands on others' limited cognition and capture their attention more often than those who refrained from launching similar assaults on people's attention. Unfortunately, the project managers in our study eventually realized what you and I have come to learn the hard way: Over time, all their extra communicating added to the amount of data that those they were trying to reach had to process, leading them to feel more distracted and making it even harder to capture their attention.

The predicament that most people find themselves in today is now well-documented. Our attention is fragmented. We lose focus and become easily distracted. We can't get things done as quickly or as easily as we should be able to, and often quality suffers. More insidiously, the multiple demands on our attention shape not only whether we can pay attention but also how we pay attention. As Nicholas Carr eloquently described in *The Shallows: What the Internet Is Doing to Our Brains*, one of the first books to document how the internet is changing patterns of attention: "Media aren't just channels of information. They supply the stuff of thought, but they also shape the process of thought. And what the Net seems to be doing is chipping away my capacity for concentration and contemplation. Whether I'm online or not, my mind now expects to take in information the way the Net distributes it: in a swiftly moving stream of particles. Once I was a scuba diver in the sea of words. Now I zip along the surface like a guy on a Jet Ski."

Since Carr published his important book fifteen years ago, our attention has undoubtedly become more fragmented as more companies and more people try to find ways to capture it. Several important books have appeared in the last few years to help us figure out how to redirect our attention. Cal Newport, author of *Digital Minimalism*, exhorts us to dampen the assaults on our attention so we can do "deep work," and Johann Hari, author of *Stolen Focus*, urges us to find ways to reclaim our focus so we can "think deeply again." In this chapter, I take a slightly different approach. Rather than tell you that you should pay better attention so that you are not so exhausted, I'm going to show you how paying attention is itself exhausting.

HOW WE PAY ATTENTION MATTERS

Maya's preshower routine is familiar to most of us. According to a major national study conducted in 2010, 65 percent of American adults reported that they routinely slept with their smartphone on or right next to their beds. By 2023, researchers had stopped asking if people slept with their phones *next* to the bed and were instead asking if they slept with them *in* bed. Sixty percent of adults admitted to doing so, and a whopping 89 percent reported that they checked their phones within the first ten minutes of waking up. Were Maya and the rest of us just glancing at the time or waking up to read one article on our phone, these stats might seem unremarkable. The real cause for concern is that we mercilessly divide our attention across so many different and incommensurate information inputs that we are literally fatiguing our brains from the moment we wake up.

A long line of research in brain science shows that the human brain is not made to do multiple things at once. When we are confronted with a new piece of information or start some new task, blood rushes to our anterior prefrontal cortex, which is like our brain's command center. This command center sounds a two-tone alarm that we need to switch our attention: The first begins a search process for neurons capable of making sense of the information or doing the task. The second sends a command that will tell those neurons what to do. This process is called "task rule activation," and it only takes a few tenths of a second to unfold. When some new stimulus demands our attention, the brain must disengage from the first task—reading the news—to begin the second—checking our text messages. The command center tells the brain that another attention switch is going to happen and sends another two-tone alarm, starting another rule activation protocol. What's remarkable is the consistency with which this process unfolds. This process occurs in the same sequence every time we switch our attention from one input to another.

Here's why that matters. Each time we are exposed to a new stimulus and engage in the sequential process of disconnecting and reallocating our attention, we are expending valuable metabolic energy. Our brain makes up around 2 percent of our body mass but burns roughly 20 percent of the calories we ingest daily. Most of that energy expenditure is devoted to controlling our bodies; the brain reserves very little energy to power its own active cognition. As neurologist Richard Cytowic notes, "Precisely because it is so efficient, [our brain's] reserve margins are slim and quickly eaten up by the demands made by constantly shifting attention."

Each switch in attention is a costly energy expenditure. And the exhaustion from each successive switch adds up cumulatively. Think about running short sprints. You run as fast as you can for ten seconds, you cover 200 feet, and you feel tired. But you have enough energy to do it again. You rest for a minute, then sprint for ten more seconds, but this time you only cover 180 feet. Of course, the longer you rest between sprints, the more the results of your second attempt will look like your first. But if you don't rest at all, you may only be able to cover 100 feet on your second attempt. That's exactly what happens in our brains. The more we switch our attention, the harder it becomes to attend to anything because the brain grows increasingly tired. After too much switching in too short a period, it's hard to sustain attention at all because our brain is exhausted. Since it uses only a small reserve of its entire energy for cognitive processing, the rest of our body might not feel exhausted, but our brain is worn out. And that's a feeling that impacts us deeply.

Cognitive scientists give these dramatic shifts in our attention the name "context switching." When we need to switch contexts, our brain must disengage from the cognitive processes related to the initial task and reorient itself to the demands of the next task. If we switch from looking at one TikTok video about cute hairstyles to another about fashionable shoes, we don't actually shift our attention in the technical sense of activating the command center in our prefrontal cortex and getting new neurons firing. It's when we switch between contexts that demand new and different kinds of cognitive processing to make sense of information or perform some operation that the process of disconnecting and reconnecting discussed earlier occurs and our brain's limited energy is discharged. Context switching also releases cortisol to give a burst of energy to power through the transition. That cortisol release comes with a nasty side effect: Short energy bursts from the stress hormone accumulate into cognitive exhaustion. Not all context switches are the same, and thus the different ways that we shift our attention exhaust us differently. I find there are three types of context switches that are the most common and the biggest taxers on our cognitive energy reserves: switches between modalities, domains, and arenas.

WHY DO YOU HAVE SO MANY APPS OPEN?

A modality is a particular mode in which something exists or is experienced. At the most basic level, our senses—seeing, hearing, touching, tasting, and smelling—are modalities. If we could isolate our senses from one another, we would experience a stimulus, like a car passing in front of us while we're standing at a crosswalk, differently depending on which sense is activated. Once, I was standing at a crosswalk and some dust flew into my eye. While my eyes were closed for the fifteen seconds or so it took to rub them and coax the fleck of dust out, I could hear a car engine. When I opened my eyes, I was surprised to see that the car was right in front of me as I stood still in the crosswalk. Those few seconds of transition from sound to sight were jarring because I had to reorient to the world when perceiving it through a new modality. It wasn't hard to do, but it also wasn't automatic. Adjusting to experience a stimulus in different ways—via different modalities— takes work.

Devices—laptops, cell phones, smartwatches—are modalities. Software applications like LinkedIn, Facebook, Zillow, Zoom, and Microsoft Word are modalities too. Each device or application gives us different capabilities. They also look different and are formatted differently from one another. These differences matter. Ioana, who works as an educational consultant for K–12 schools, is often on several videoconferences a day with various school administrators around the country. The schools typically arrange the meetings with Ioana, and she uses their preferred videoconferencing platforms. In a single morning, she often has meetings with schools through Zoom, Microsoft Teams, Webex, and Google Meet. As Ioana recounts, "We primarily use Zoom at my company, so I'm the most comfortable with it. When I get on Teams or Webex, I'm just not as fluent. It takes me a minute to think about how to share my screen or change my background. You do those things differently on each application, so I get a bit stressed in the moment when I can't just do it as easily as I could on Zoom." Although it barely seems like a notable switch to move between applications of a similar type like Zoom and Teams, our experience on each of those platforms is unique because we need to readjust to their layouts, colors, and buttons to be able to use them effectively. That readjustment is enough to cause us stress, and the cognitive expenditure and cortisol release of moving between different types of applications—from Gmail to your company's financial management platform to the ParentSquare application that sends notifications about your child's school—is even greater.

These modality switches are increasingly common in our typical days. In one recent study, investigators examined twenty teams, consisting of 137 individuals, across three Fortune 500 companies for five weeks. Most of the teams worked back-office jobs in areas like finance, inventory management, and supply chain operations. Using data logs from their computers, the researchers tracked how often they switched between modalities in each workday. They found that the average person moved between different modalities—here, apps or websites—nearly 1,200 times each day. The data logs showed that, on average, it took two seconds to switch between modalities. If you add that up, it means these workers spent over three hours a week switching between modalities.

I recently conducted a survey together with colleagues at Asana and Amazon of 3,000+ knowledge workers in the US and UK to learn more about how context switching between modalities works. We found that people spent, on average, 57 minutes per day (4.75 hours per week) switching between software applications and websites while at work. That number didn't surprise us. But we were surprised to learn that they also reported spending 30 minutes a day *deciding* what application they should use for a specific task. Do I ask for this favor on email or Slack? Do I make this table in Word or PowerPoint? Do I look up this definition on Dictionary.com or do I ask ChatGPT? None of these decisions seem overly complicated or consequential, and they're not. But they do take some investment of thought and action. If you have to make these micro-decisions about which modality to use each time you switch, and you're switching 1,200 times a day, it's easy to see how the accumulation of decision fatigue can lead to exhaustion. There is mounting evidence that the more often we have to make small decisions, the more exhausted we feel. We need to factor the invisible labor of context switching into our understanding of attentional exhaustion.

Before the blood even flows to the prefrontal cortex for an actual attention shift, we've already been making micro-decisions about which modality is best for a given purpose and what the consequences of a poor choice could be. Whenever my teenage daughter grabs my phone, her eye roll says "You're such an old guy." Her mouth says "Why do you have so many apps open?" and she proceeds to shut them all, reminding me that my phone runs slower and wears its battery out faster when so many applications are active. I suppose that's a good metaphor for what happens to us when we switch so frequently between apps and across devices. Each switch in modality by itself is relatively inexpensive in terms of energy expenditure. But the fatigue and stress associated with each switch accumulates rapidly throughout the day and over weeks in ways that lead to Level 2 exhaustion.

SHOOT! WHERE WAS I?

A domain is a specific set of tasks or activities that requires a specific type of cognitive processing. Each task or activity that we do imposes different demands on the brain's cognitive resources. Giving someone advice about how to deal with a difficult colleague is a more abstract, empathetic, and possibly emotionally engaging activity. It primarily involves social cognition, understanding human emotions, and communication, activating regions of the brain like the temporal lobes (for language comprehension) and the limbic system (for emotional processing). By contrast, preparing a financial report for your boss is a highly analytical task that requires attention to detail, numerical analysis, and logical structuring of information. These requirements are typically met by involving the dorsolateral prefrontal cortex (for problem-solving and attention) and the parietal lobes (for spatial and numerical reasoning). Switching tasks across domains thus requires a shift in activation from one set of neural networks to another, which is not instantaneous and requires significant metabolic energy. Switching between tasks or activities in different domains requires the brain to reallocate cognitive resources and adjust its processing strategy.

We switch domains often throughout the day. That's because the roles we play both at home and at work involve many different kinds of tasks. What typically separates high performers from others is that they can quickly switch across domains without much lag time. For a reason that science has yet to fully figure out, some people's brains are able to shift between neurons faster than others and do so with less residue. But that's not the norm. For most of us, switching our attention from one domain to the next is a tiresome act. Although humans have been making such switches for hundreds of thousands of years, evolution did not prepare us to make them with the speed and frequency we do today. "That switching comes with a biological cost that ends up making us feel tired much more quickly than if we sustain attention on one thing," observes behavioral neuroscientist Daniel Levitin. The multiple devices we carry and the applications we have open on them make us available to be pulled into different domains with alarming speed—and often while we're still in the middle of focusing on something else.

The major problem with domain switching is that although our senses may be able to move quickly from giving advice to working on a financial report, our attention moves much slower and is more reluctant to stop doing what it was doing. Vicente, a high school science teacher and track coach, runs into this problem a lot. A gregarious and generous coach, Vicente encourages students to message him on the team's group messaging app if they have issues or just want to talk. "You'd be surprised at all the emotions these kids have just making it through the day," he says. "Sometimes they just want to tell you a few things that happen. I get it. I really try to listen and I feel for them." But Vicente finds that messages seem to arrive at random times, like when he's grading assignments, returning emails, playing video games, or helping his own children with their homework. "If I'm trying to pay my bills in my banking app and I get a message from one of the kids on the team, I usually read it. But then when I go back to doing my banking stuff, I make a lot of mistakes because I'm still thinking about the kid I was just messaging with. That's not so good. Sometimes I just start all over because I lost my place and can't get back into it."

The problem Vicente experiences is known as "attention residue," and it occurs when part of our attention is focused on a task in another domain instead of being fully devoted to the task at hand. As Sophie Leroy, a management professor at the University of Washington, notes, "Attention residue easily occurs when we leave tasks unfinished, when we get interrupted, or when we anticipate that once we have a chance to get to the unfinished or pending work we will have to rush to get it done. Our brain finds it hard to let go of these tasks, and instead keeps them active in the back of our mind, even when we are trying to focus on and perform other tasks." Gloria Mark, a professor at the University of California, Irvine, has probably done more research than anyone on the effects of attention switching across domains. She likens Vicente's mind, and ours, to a whiteboard: "Just like sometimes you can't erase that whiteboard completely: you see traces of what was written on it. Same thing happens in our minds, and that residue can interfere with our current task at hand."

Mark started studying people's attention span on their devices and applications back in 2003. In her early studies, she found that people averaged spending about two and a half minutes on any screen before switching. In 2012, she and her colleagues found that they averaged 75 seconds on any screen before switching. And in 2016, the last time she checked in a formal study, she found that they averaged 47 seconds. Mark would be the first to tell you that switching between modalities is costly enough if people do so while working on a task or activity in one domain. But it is much more damaging if those modality switches are accompanied by switches in domain. To investigate this issue, Mark revisited her data and only counted a switch in modality when it also accompanied a switch in domain. When looking at the data in this way, she found that people typically spent about ten and a half minutes on a task or project in any domain before switching, but when they switched back and forth, it took twenty-five and a half minutes, on average, to get back on track and return to the same level of focus they had before shifting their attention.

Mark isn't the only one whose data show the costs of attention residue on productivity and exhaustion. A recent survey of knowledge workers in the US and UK found that 43 percent of respondents directly connected their exhaustion to switching between digital tools. Interestingly, although study respondents reported that domain context switches exhausted them, they also felt they were necessary to be productive. As Mark describes, "We may have the illusion that we are doing more and that our human capacity has expanded when we shift our attention, or multitask, but actually we are doing less." The effects of domain switching are so pervasive that people don't even have to actually switch domains to produce the brain fatigue associated with a switch. One study found that simply thinking they might have to make a switch soon reduced test takers' performance by 20 percent.

Vicente knows that as soon as he starts playing *Call of Duty* on his Xbox gaming console, he could get a message from one of his athletes about what distance to run over the weekend. "Sometimes I think about turning off notifications when I'm gaming because it kind of makes me nervous to get interrupted, and then it takes me a minute to get back into the game," he says. "But I don't ever do it because I want to be there if they need me."

HONEY I'M HOME—WAIT, I NEED TO TAKE THIS WORK CALL

My honeymoon was also a work trip. My wife and I arrived on the island of Kaua'i and spent a week lying on the beach, taking beautiful hikes, and eating delicious food. Then we flew to O'ahu where I had a work conference. We were there for four days. I skipped out on many of the conference sessions to continue my honeymoon, but I remember many instances of being distracted reading text messages on my phone from colleagues who were attending. This was before the era of smartphones. I can only imagine how annoying I would have been if I had one of those back then. The fact that I was at work on my honeymoon sounds terrible when I read it on the page. But even now, after spending twenty years studying and thinking about digital exhaustion—and having seen all the evidence about how detrimental it is to our well-being to switch quickly between distinct life arenas such as work and home—I still think it was a smart move to get work to cover part of my honeymoon. The fact that I continue to think this way means that the culture of acceptance around disintegrating work-life boundaries is strong, at least in the Western industrialized world.

Some of the earliest people to experience the rapid and jarring attention switches between work and home made possible by digital technologies were, as we used to call them, teleworkers. The term "telework" was invented in the 1970s to describe knowledge workers who connected to the office from home via a telephone line. Although we've talked about telework for half a century, it wasn't until the diffusion of broadband internet in the mid-2000s that telework became a real possibility for the average worker. To explore how digital technologies changed when and how people worked in the aughts, Michele Jackson, a communication professor at the University of Colorado, and I launched a study of teleworkers across a variety of industries. Most worked from home a few days a week and in the office the other days, but some worked from home full-time. Such telework arrangements were not common then, and these folks were on the vanguard.

Invariably, the people we interviewed in the early years of the 2000s discussed the advantages of their work arrangements. Cynthia, a sales representative for a major telecommunications firm, told us, "I love teleworking. It's amazing to be able to do a sales call, then quickly run out to the bank. Or if I'm working on a report, my daughter can call me from her cell phone, and I can answer without feeling like I shouldn't be taking a personal call at work." But in every interview, usually several exchanges after first discussing their jobs, Cynthia and the other teleworkers made a confession: "If I'm really honest, it's hard to switch between work and home. It's exhausting because I get invested in something and then I've got to make this big switch between two worlds. And it happens all the time now."

Flash forward to the year 2022. My research team was conducting a study of knowledge sharing at a Fortune 100 network device company where employees had the option to work remotely up to three days a week. Ian, a software engineer who took the maximum three days, made an observation about his arena switching that sounded exactly like Cynthia's nearly two decades earlier: "Working remotely has its problems for sure, but overall, it's been pretty awesome. It's great to be able to just work when you need to but then switch to focus on doing family-related things on the fly. I get notices on the SportsEngine app for my daughter's club soccer team, and I can immediately check them and then call the coach or message other parents to coordinate a carpool. It's amazing to be able to do that while I'm working." But just like Cynthia, Ian's optimism about arena switching also belied a bigger problem. As he later recounted in a vulnerable moment, "It's kind of hard to shift your attention from work to your home life and back again all the time. It kind of wears me out because I get worked up about something that's going on at my kid's school and then it's tough to just go back and have a meeting with my team right after. I'd say I'm definitely just getting exhausted by all the back-and-forth."

Our digital devices and all the applications on them make switching between arenas feel effortless. Messages from our apps punctuate our work meetings or concentrated work time, and emails, Slack messages, and phone calls interrupt our time at home or with family and friends. So many people have admitted to me—sheepishly, I would add—how they've stolen away to the bathroom while they were out to dinner in the evening with friends to sneak a peek at their work email or to return a Teams message they didn't get to earlier. But that sheepishness is often accompanied by a sense of justification that the flexibility their digital technologies provide them means that they don't have to make such fine-grained distinctions between the arenas of work and home, and that they can engage in even *more* social or personal activities because their work is so flexible. If they were really bold, they might even go so far as to say that they could afford to spend more time on their honeymoon because they could keep up with their conference while lying on the beach. But would someone really go that far?

One of the most penetrating insights into the exhaustion that arises from switching our attention back and forth across the arenas of work and home can be found in Christine Beckman and Melissa Mazmanian's wonderful book *Dreams of the Overworked*. Beckman and Mazmanian spent three years exploring the lives of nine professional families in California with young children. The authors conducted detailed observations of the parents at work and also participated in the families' home life, resulting in an intimate and nuanced understanding of the challenges and dynamics these families face as they navigate work, parenting, and personal well-being in the digital age. The book is filled with many heart-wrenching stories of people on the brink, so exhausted by trying to cope with the overlapping demands of work and home, as well as the unrealistic expectations that society places on them to be perfect in both arenas. The authors conclude that the major issue facing the modern professional family is that our digital technologies collapse the distinction between work and home. Both arenas perpetuate a culture of constant availability that exacerbates feelings of stress and exhaustion—work colleagues expect to reach you at any time, and so do your spouse and children. What's more, Beckman and Mazmanian show that the working professionals they followed coped with the demands of constant availability by using their digital devices to coordinate—often on the fly—support from extended family, neighbors, and professional caregivers so they could continue to manage their multiple responsibilities. But these hidden support structures came at a cost. The reliance on the people who helped them resulted in more texts, emails, and phone calls that shot more holes in the boundary between work and home.

Beckman and Mazmanian's research illustrates how our digital technologies create an autonomy paradox. Although our devices can enable us to perform tasks anytime and from anywhere, they also lead us to feel compelled to be always available and responsive. This constant connectivity blurs the boundaries between work and personal life, leading to stress, exhaustion, and overall drops in reported well-being. The very devices that offer freedom can also make us feel trapped in a perpetual cycle of work.

The growing sentiment among people I've studied over the last twenty years is that it is inevitable that the arenas of work and home will collide in this era of digital connectivity. They're probably right, although as I write this, lawmakers are drafting proposals that attempt to place limits on how much companies can intrude on people's home lives. In February 2024, the Australian Senate approved a bill that allows employees to disregard work-related calls and messages outside of their scheduled hours without any repercussions. This initiative mirrors France's 2017 legislation that established workers' right to disconnect after hours, a policy that Germany, Italy, and Belgium have also adopted. Inspired by Australia's attempt, California

Assemblymember Matt Haney announced Assembly Bill 2751 in April 2024, which would require a public or private employer to "establish a workplace policy that provides employees the right to disconnect from communications from the employer during nonworking hours, except as specified." Haney argues that Californians need such protections from their employers because "smartphones have blurred the boundaries between work and home life. . . . People have to be able to spend time with their families without being constantly interrupted at the dinner table or their kids' birthday party, worried about their phones and responding to work."

Such proposals are certainly well-meaning. The thousands of people I've talked to about digital exhaustion over the years would likely welcome the "right to disconnect." But most would find it unfeasible to do so even with laws in place. Work and home are so thoroughly joined by digital technologies that it is nearly impossible to keep them apart. A recent study by University of Pennsylvania Wharton School professor Nancy Rothbard and colleagues shows how much social media platforms like Facebook and LinkedIn contribute to blurring the boundary between work and home. They found that two-thirds of working adults on Facebook were connected with colleagues on the platform. The threads of friends, coworkers, and family members were all interwoven in one place, forcing quick and repeated attention switches between information coming from work and home. Participants had more difficulty maintaining a boundary between their work and home lives when people from each arena shared space on their social media platform. They also found it difficult to reject friend requests, especially from work colleagues—which, of course, brought the two arenas into even closer contact.

"Right to disconnect" laws might stop a specific work email from entering the arena of home after 5 p.m., but they can't sever the overlaps between work and home that are so thoroughly intertwined in the broader set of digital technologies we use every day.

And as we've seen in this chapter, our exhaustion doesn't merely come from too much connection. Rather, it comes from the demands of switching our attention rapidly and consistently across the various modalities, domains, and arenas that make up our modern lives.

CHAPTER 2

Inference:
Traps in Every Direction

We yearn to know what motivates others and drives their actions. It's not just because we're curious; it's also because understanding why people do what they do and act the way they act helps us figure out how we can best interact with them. We use data we can observe about people—what they say, how they say it, where they go—to interpret and make inferences about their motives. For much of human history, we have been limited in the data available to us about people we don't see or interact with regularly. But in the digital age, the amount of data about others that is available at our fingertips is almost limitless. We now can and do make inferences about almost everyone we have exposure to online, whether we know them or not. You might think that more data about people's behaviors would make understanding them easier. In this chapter, I'll show that it actually makes it much harder and more exhausting.

Take Aaliyah as one example. Aaliyah works as the executive director of a large NGO in the state of Georgia. She uses a variety of digital technologies to communicate with coworkers, stakeholders, and community members about inequality and justice issues like gay rights, abortion, and racial discrimination. It's demanding work, and Aaliyah is always concerned about making sure people feel comfortable, understood, and represented in the right ways. She finds that the digital footprints people leave online make it easy for her to make inferences about their motives and behaviors. "I just can't help myself," she observes. "There's so much stuff out there that people post that I find it easy to start to believe I know what they're thinking or feeling when I don't. It's also super easy for me to start to think about what they think about me and for me to start thinking about who I really am when I respond to them. It's all really exhausting."

Aaliyah's remarks about the nature of observation and inference in the digital world reflect three insights we'll explore in this chapter. The first is that digital technologies act as prisms. Prisms reflect light, but they also distort it. Online connectivity gives us unprecedented access to people's actions and opinions, and we use those data points to make inferences about their motivations and desires. The more data points we see, the more likely we are to make assumptions about people and forget that they are just that—our own inferences about why people act, not direct windows into the motivations for those actions. I'll show evidence that it is exhausting to draw so many conclusions about people on a regular basis. But as astute observers like Aaliyah know, it's also exhausting to try to stop yourself from believing the inferences that you do make.

The second insight buried in her observation is that digital technologies act as portals through which we can really get to know what someone is thinking and why they think it. Because most digital platforms give us access to the historical back-and-forth of our conversations, it is easier than ever to put ourselves into other people's shoes to see how what we did or said made them think, feel, or act. In short, it's increasingly simple to see what inferences others make about us and then to make inferences about why they made those inferences about us. Are you tired from the mental gymnastics required to decode the prior sentence? If so, you are starting to get a sense for how using digital technologies as a portal into other people's mental states can and does increase our own exhaustion.

A third insight we can learn from Aaliyah is that digital technologies act as a mirror, reflecting back to us the many things that we've done and said. Our constant exposure to our past activity often causes us to make inferences about ourselves. A great many of the people I interviewed for this book said things like: "When I scroll through my Facebook feed, I sometimes think I'm not that nice of a person." Or "When I see how often I've written 'sorry for the delayed response' it makes me wonder whether I'm as helpful as I think I am." Our repeated and sustained interaction with our past selves is a major source of exhaustion because reappraising and reckoning with who we are is one of the most mentally taxing and emotionally exhausting things we do with our digital tools.

HOW ELECTRONIC EXPOSURE DISTORTS REALITY

Dean and his college friends spent six months planning a two-week bike-riding trip through France and Italy. Dean was super excited to go. But as the date approached to buy airline tickets and make hotel reservations, he faced the sobering reality that he just didn't have enough money. It was a tough decision, but he had been working to be more financially responsible and spending so much money when the balance in his savings account was so low just didn't seem like the right choice. As his friends departed for Avignon, he texted them well-wishes and told them to make sure they posted lots of pictures on Instagram.

During his friends' trip, Dean monitored his Instagram feed and saw photos of them riding through the picturesque French countryside, drinking beers at local pubs, and playing bocce ball with locals in beautiful Italian parks. As Dean recounted, "It was hard to see those pics of their trip. They were having the time of their lives, and I was just stuck at home working. I could have been there." As the trip progressed, the group posted more and more pictures to Instagram and sent Dean the occasional text telling him that they wished he were there with them. Dean looked at his feed with envy. But he really hit a low point when he saw pictures posted by his ex-girlfriend, who happened to be in Europe at the same time, eating at cute Viennese cafes with a guy who appeared to be her new boyfriend. "It was a shitty summer," Dean later recalled. "Everyone was out living their best lives except for me. I felt pretty depressed."

Dean's story reminded me of a series of studies I've always loved, which also involve biking and Europe. Terence Mitchell and a team of researchers at the University of Washington gave undergraduate students taking a three-week bicycle trip across California a variety of assessments before, during, and after the trip. They were asked to rate things like their expected and actual enjoyment of the trip and how difficult it would be. The study aimed to see how their actual experience of enjoyment compared to their expectations of the trip before they left and their recollection once it had concluded. The authors found that their subjects' prospective feelings of enjoyment were much higher than their actual enjoyment of the trip (measured while they were on the trip). But their ratings of enjoyment were highest *after* the trip was over. As the authors concluded, "Disappointment, while acute during the actual experience, is short-lived."

The same author team conducted a similar study of people who took a guided twelve-day European tour, evaluating people's enjoyment of the trip before, during, and after it occurred. And they found the same patterns as they did in the bicycle trip study: People's actual enjoyment of the trip was much lower than their memory of enjoyment on the trip three weeks later. Reflecting on the results of these two studies, the authors argued that people often take a "rosy view" of their life experiences. That is, they anticipate events will be better than they actually are and remember them more fondly than the way they experienced them at the time.

I wondered if Dean's friends who took a bicycle trip through Europe would have felt the same way, so he introduced me to two of them. When I asked them to share some of their favorite memories from the trip—which had now occurred a full three years ago—they both recounted similar experiences. They described the beautiful farms in the south of France and recounted how fun it was to go to local pubs. One of the friends told me about a special memory of a time when they played bocce ball with some old Italians in Northern Italy—and lost badly! Those stories sounded familiar to me because they were the same ones Dean described seeing pictures of while he was stuck at home. I asked both friends if they had any of the text messages they'd sent during that time, but neither could find any on their phones given how long ago the trip occurred. One of the friends told me he had been texting with his older brother a lot, so I asked if I could talk with him. The older brother's name was Zach, and when I inquired about what he remembered about his brother's bicycle trip through Europe, he laughed and said, "I'm so glad I didn't go on the trip. He texted me all the time complaining about stuff like how there was too much rain and his bike kept breaking and how the guys were annoying him. Oh, and he was always going off about how much the beer sucked compared to home. No IPAs anywhere, apparently. It sounded like he was miserable."

Mitchell and his colleagues conducted their studies a decade before people carried high-resolution cameras in their pockets and could easily post pictures on social media for all their friends and the whole world to see. A more recent study found that participants were more likely to remember an event favorably *after* it occurred if they took a photo of it than if they did not. The authors also found that taking a photo of an experience improved people's evaluation of the experience *during* the experience. The design of these experiments ruled out the possibility that the findings were a result of people simply being more likely to take pictures of experiences they enjoyed. Instead, the data showed that people who took pictures of certain experiences were more engaged in the experience and were thus better able to focus on and remember them. Another study of adult Instagram users showed that people who posted photos of an experience viewed the experience as more favorable than people who took pictures but did not post them. When we look across these related studies, a clear pattern comes into view: 1) People's memories of experiences are generally more favorable than the actual experience itself, 2) they will view the experience as even more favorable if they took a picture of it, and 3) their favorable view increases if they posted that picture for the world to see. Dean's friends might not have had an amazing experience while on their bike trip through Europe, but they did all the right activities to assure that they would remember their experience as amazing.

That all sounds great for Dean's friends. But what about for Dean? For years, he assumed he'd missed out on an epic trip. He saw his friends' photos posted to Instagram and inferred that they were having the time of their lives. The evidence seems to suggest that they were not, indeed, having the time of their lives while the trip was in progress. But their memories of the trip—aided by the fact that they took and posted photos about it that linger on social media to this day—led them to remember that they were. This is perhaps the major conundrum of social media and all digital technologies that connect us to information about others. The data we see are curated. People record their most favorable experiences or outcomes (which are made more favorable in their recording), and their natural tendency toward rosy retrospection is intensified by their ability to continually relive those positive experiences through pictures and other digital documentation that persists over time. The salience of those positive experiences makes it easy to engage in the natural tendency to forget or downplay all the negative experiences that were not similarly captured and preserved. Who's to say that Dean's friends didn't really have the time of their lives if they only remember having the time of their lives? Does it matter what *actually* occurred or only what we *think* occurred?

What we can say for sure is that it matters for Dean, and it matters for you and me. One of the most robust areas of research into social media concerns the topic of social comparison. In his seminal research on the topic in the 1950s, Stanford psychology professor Leon Festinger showed that people just can't help but compare themselves with others. Festinger argued that the most common and emotionally exhausting form of social comparison is "upward comparison"—the act of evaluating ourselves against those we perceive to be superior in some way. This process can either motivate us to improve or lead to feelings of inadequacy, depending on the context of the comparison and the perceived gap in abilities or qualities between us and others. As you might imagine, social media is a breeding ground for social comparison. The cumulative findings on this topic point in a unified direction: People who have exposure to data about others on social media tend to engage in upward comparisons that make them feel demotivated, dispirited, anxious, and hopeless, and that reduce their feelings of self-esteem—all symptoms associated with exhaustion.

A major culprit of such social comparison behaviors is our tendency to make inferences about others. When Dean saw the pictures of his friends on their bike trip, he inferred that the photos depicted an average day for them. He inferred that they were having an amazing time. *And* he inferred, via social comparison, that his day, or his summer, or his life, was not as good as theirs. Dean may have had good reason to make such inferences. But the data from which he was building them were incomplete. They lacked nuance and insight about the situations that led to them or the hardships that preceded and followed. They represented a distortion, from which Dean constructed a reality.

This link between inference and social comparison is not confined by any means to social media, although we see its effects most pronounced among users of such platforms. Exposure to data about what projects people have worked on or are currently working on from our company's Slack channels or data about coworkers' sales revenue for the quarter are examples of how the recording and persistence of data creates the conditions for inference-making via social comparison. While Dean's experience highlights how the data we see hide deeper truths, we often fail to recognize or forget that we can only ever see part of the picture.

HOW WE THINK WE'VE GOTTEN INSIDE OTHER PEOPLE'S HEADS

A few years ago, senior leaders of a big Peru-based telecommunications company approached me because they were interested in improving knowledge sharing across various divisions. They reasoned that if people in one division who came up with successful solutions to problems shared them with folks in other business units the whole company could benefit. But people in different business units didn't talk with one another very much, largely because they didn't work together and were located in different buildings, cities, or countries. They wondered if using a social networking technology for businesses like Salesforce's Chatter might help them to connect and, ultimately, share more knowledge and information with one another.

So Samantha Keppler, who is now a professor at the University of Michigan, and I ran a series of studies to explore how they might use Chatter most effectively. We found that people were more likely to ask a coworker for work-related knowledge if they had previously communicated with that coworker about non-work matters. For example, someone's propensity to ask a coworker they didn't know for knowledge or advice about financial reporting increased if they'd previously had a conversation on Chatter about soccer. We called this effect "social lubrication," meaning that work-related knowledge moved out of the silos in which it was stuck when it was lubricated by social interaction. These findings led the company's senior leadership to appreciate the role that communication about non-work topics could play in spreading knowledge across the company, and they changed their policy from prohibiting such communication to encouraging it. That was good news.

But when we dug deeper into the data, we found a puzzling trend: A lot of people began asking for knowledge from others with whom they had previously never communicated directly. The data logs showed that they had never DMed each other on the platform or emailed each other, and we knew that it was unlikely that they'd seen each other in person since most worked in different cities. We couldn't understand why people's willingness to ask others for knowledge increased among this sample of individuals who hadn't discussed either work or non-work topics together at all. We focused our attention on the instances in which these people decided they needed knowledge from a particular person but waited a significant period of time to ask them for it—usually between three weeks and three months. Then we examined what they did during this time. We found that in between deciding they wanted to ask someone for information and actually asking them, these inviduals began to interact with the coworker *indi-rectly* on Chatter. They'd put likes next to the coworker's posts or comments in their thread. They might even @mention the coworker in one of their own posts if it described something they thought the other person might find useful or novel. Interestingly, we observed very few instances of reciprocation. It was like these people were having a bunch of interactions with a rock. They were doing all these things to show their coworkers they were seeing them and thinking about them, but the coworkers almost never responded.

Surprisingly, that reciprocation didn't matter. Engaging in these anticipatory communication activities was enough to make these knowledge seekers feel confident that they had a sufficiently strong relationship to burden someone with a big ask. As a customer support technician named Sandra told us, "Oh, he knows who I am and he likes me because I'm always commenting on his posts and mentioning him in mine. When I look back at my post history you can see that we've had a lot of engagement together." I'm still not sure exactly what "a lot of engagement" meant, because the person she mentioned "engaging" with never once responded to her posts or liked anything she posted. But we heard similar stories from other workers. It seemed that the more they made public and visible overtures toward someone, the more they became convinced that the other person knew and liked them. And our results showed that the more positive their inferences were about the other person's feelings toward them, the more likely they were to ask them for knowledge or information. The data showed a direct link between the frequency with which they created signals of a relationship with a specific person on Chatter and their likelihood of reaching out to that person, even when the knowledge holder reported that they did not know the knowledge seeker.

What I've always thought was powerful about that finding is how confident someone could be in thinking they knew what was going on in someone else's head. Ultimately, it didn't matter that the person Sandra was pursuing reported not knowing her at all. From Sandra's point of view there was a relationship there. Sandra used her own trail of activities on Chatter to infer that the other person knew and liked her.

Behavioral cues made visible through digital technologies lead to diverse interpretations regarding peoples' motivations, goals, and desires. We humans naturally make such inferences based on observable actions. What we often don't consider is the toll that making so many inferences takes on us. In their classic book on social cognition, psychologists Susan Fiske and Shelley Taylor argued that humans use two primary modes of information processing when forming inferences about other people's behaviors and motivations. These two modes, often referred to as automatic and controlled processes, represent distinct ways our brains function to handle the complexity of social interactions. In the automatic mode, our brains operate quickly, with little effort or conscious control. We draw on our understanding of traits or established patterns to make swift judgments about others, which can be essential for navigating social situations efficiently. These judgments are typically based on minimal information and can be influenced by stereotypes or previous experiences. In contrast, controlled processing is slow, deliberate, and effortful, requiring conscious attention and cognitive resources. This mode is engaged when we face novel situations or when a decision requires more thought, such as understanding complex social cues or interpreting mixed signals. Controlled processing allows for a more thoughtful consideration of the social environment and can override automatic responses that might be inappropriate or incorrect.

As you might imagine, automatic processing is fairly effortless, while controlled processing puts the brain into overdrive, demanding a huge supply of cognitive resources to make sense of ambiguous stimuli and encode them into categories we can understand. Fiske and Taylor found that people often try to avoid the effort associated with controlled processing by willingly reverting to the stereotypes, schemas, and other shortcuts that characterize automatic processing. They observed that people revert to such heuristics so often that they like to call humans "cognitive misers." That's the mental component of exhaustion.

Controlled processing is also linked to emotional exhaustion. When we go through the demanding work of decoding a stimulus and making an inference about someone's motivations or mental state, we reflect on that inference in a way that we don't when we make inferences through automatic processing. Being hit with the realization that you understand someone's motives is jarring and takes its toll. A 2018 study that examined how college students on Facebook responded to ambiguous posts showed that study participants felt emotionally exhausted after making inferences about other people's mental states and motivations, such as whether they were depressed or were simply posting something to gain sympathy or attention. Arriving at these conclusions after piecing together evidence from multiple posts was mentally exhausting. But stepping back and realizing what they thought the person was trying to do by posting certain content elicited emotional responses that left them feeling depleted and dispirited.

WHY WE CAN'T STOP LOOKING AT AND EVALUATING OURSELVES

During the COVID-19 lockdown, Graham found himself feeling increasingly unsatisfied with his looks. He'd always thought of himself as someone with good self-esteem and a healthy attitude toward his body, but things had changed. He began to notice that his skin was blotchy, and he wondered if he was gaining weight because it seemed he had more flab under his chin than he remembered having in the past. And he no longer liked the shape of his nose. "Why am I thinking about this stuff? This isn't me!" Graham lamented.

Graham's experience of increased self-criticism is, unfortunately, common among heavy users of video-based digital communication tools like Zoom, FaceTime, or Microsoft Teams. We're not accustomed to seeing ourselves for multiple hours a day. But that "self-view" window makes it difficult for us not to look at ourselves. People spend proportionately more time looking at themselves during a Zoom call than they spend looking at the other people in the meeting. And the research is quite convincing that the more we allocate our attention to our physical appearance, the more worried and stressed we become about the way we look.

Jeremy Bailenson was among the first researchers to systematically study Zoom fatigue—that feeling of exhaustion you get after being on video calls for several hours a day—during the COVID-19 pandemic. As Bailenson observed, the fact that we regularly see ourselves during interactions via the self-view feature in most videoconferencing platforms is somewhat absurd: "Imagine in the physical workplace, for the entirety of an 8-hr workday, an assistant followed you around with a handheld mirror, and for every single task you did and every conversation you had, they made sure you could see your own face in that mirror. This sounds ridiculous, but in essence this is what happens on Zoom calls."

What makes looking at ourselves on a screen so exhausting? Research shows that seeing live video reflections of ourselves and attending to others on camera increases our cognitive load. The increase is significant because we have to manage nonverbal communications that are natural in face-to-face interactions but become laborious online. For example, if we raise our arm to indicate the size of a ball we are talking about, we recognize that our arms extend outside the camera's field of view, so we have to quickly move our hands to be visible to our communication partner on the other side of the screen. We don't usually think about these nonverbals when we communicate in person, but the intense gaze of the camera, coupled with its limited field of view, make us aware of our actions and force us to constantly adjust in response to our self-monitoring. Our cognitive load also increases because the camera only focuses on part of the environment and leaves other parts open to our imagination. We can sense this, and we attempt to compensate for it.

As Xiao, one of my students, admitted to me in office hours a day after we had a Zoom session in class: "When you're talking, I'm always taking notes. But when I looked at myself on the screen I couldn't see myself taking notes, so I thought that that means you can't see me taking notes either." She continued, "So then I was trying to nod my head more than I normally would in agreement so you could see I was following along. I could see myself doing that in the video. But then when I look at the video I see I'm not nodding, so then I thought I should start nodding so you could see me nodding. But I was wondering if you could see it too." I wasn't sure why she was telling me all this until she finally admitted, "I've been worried about it. That's why I wanted to meet with you." I was exhausted after hearing Xiao's self-reflection about seeing her self-reflection. I told her that I did notice her nodding, and then I asked her if after doing all that thinking about the version of herself she was presenting on the screen and all that work to manage it, she was able to really pay attention to and grasp the subject matter we discussed in class. She looked at me sheepishly and said, "Not really. I was too busy thinking about how I looked to you that I really didn't get what you were talking about."

Since the onset of the pandemic, there have been numerous studies linking the new reality of seeing ourselves on-screen and reflecting on how others perceive us to both mental and emotional exhaustion. So why not just turn off the self-view, you might ask? Well, first, most people just won't turn it off even though they know they probably should. There is a great deal of evidence that we can't help but attend to reflections of ourselves. It's not because we are self-absorbed but because seeing how other people see us helps us figure out who we are. So it's not surprising that most people keep self-view turned on.

The actual reflections that we see in self-view on videoconferencing platforms and even in video games are but one tiny fraction of the bigger self-view phenomenon inherent in our digital tool use. It's not just impressions of our faces that we see when we use digital tools— we leave comments on people's Instagram feeds, explanations of problems we've had in our team's Slack channels, and bad jokes in the chain of comments in our instant messages. The way we use our digital tools makes it easier to pay attention to ourselves and spend more time doing so because there is always a digital impression we can return to and see how we must look to others. We see reflections of ourselves in the digital traces we leave through every tool we use.

These traces linger long after we've inscribed them in 1s and 0s, which means we often have to reengage in the present with past presentations of ourselves. And it's here, in this temporal and spatial distance from our past selves, that emotional exhaustion often rears its ugly head. In a study led by Nicole Ellison at the University of Michigan, researchers interviewed people about their profiles on several popular online dating sites. What was unique about the study was that the researchers brought printed copies of the person's online dating profile with them to the interview. They asked the participants to rate the accuracy of each element in the profile and also made independent assessments of the person's physical attributes and traits relative to what was written in the profile. The findings showed many discrepancies between the profile statements and the offline realities of the person who created them. For example, people often portrayed themselves as thinner or more muscular in their profile than they were offline. And they would also claim that they were sportier or involved in more activities than they actually reported doing in interviews with the research team.

Participants were uncomfortable when presented with these discrepancies between the digital record of a past self-presentation in their profile and the reality of their offline persona today. But they weren't uncomfortable because they had been called out for lying. Rather, their discomfort grew from the fact that they had created those profiles with the assumption that by the time they made a match with someone on the dating site and met the person offline, their physical appearance and their lifestyle would resemble the person they conjured in the profile. But by the time of the study, most had not achieved those aims. As the authors suggested when reflecting on the interviews, the profile wasn't a lie so much as it was "a promise made to an imagined audience that future face-to-face interaction will take place with someone who does not differ fundamentally from the person represented by the profile." As they go on to say, "Our data suggest that online daters rationalized profile discrepancies by appealing to the temporal nature of promises. Specifically, participants selected attributes from a library of selves—past, present, and future—to construct a collection of identity claims that enabled them to include 'enhancements' while still self-identifying as an honest broker or promise-keeper." It is this process of rationalizing our current view of who we are with the digital traces we've left of ourselves in the past that creates heightened emotional exhaustion. Contending with the difference between who we thought we were, or who other people thought we were, and who we are now, even if those changes are positive, is an emotionally charged experience for most people.

We might be able to turn off the literal self-view on our video conferencing tools, but we can't turn off the metaphorical self-view that all of our posts, profiles, documents, emails, and comments display to us daily. Reappraising and reckoning with who we are as people are among the most mentally taxing and emotionally draining things we do in our modern world. Our digital tools certainly aren't the first mirrors in our lives, but they are the most pervasive, the easiest to access, and the most enduring.

CHAPTER 3

Emotion:
Feelings from the Screen

While attention allocation and inference-making each impact our likelihood of exhaustion, they also work together to wear us out by fostering a range of affective responses, more commonly known as emotions.

Study after study has shown that of all the contributors to burnout, the strongest is emotional exhaustion. Even positive emotions can wear us out. Strong emotions trigger physiological responses in our bodies, increasing our heart rate, blood pressure, and stress hormone production, all of which require energy. Processing and regulating emotions requires cognitive effort. Managing and controlling our emotional responses, particularly in challenging or stressful situations, is mentally and physically draining. In short, emotions are exhausting. Emma Seppälä, author of *The Happiness Track*, shows in her work that the emotions we experience tax our minds and bodies about equally. Her brain imaging research reveals that our amygdala lights up when we feel intense emotions. That's the same region that lights up when humans experience a fight-or-flight response. To combat this response, we use emotion-regulation strategies that come from a different part of our brain, located in the prefrontal cortex, to calm ourselves down. "The result?" Seppälä writes. "You tire easily. Whether you're getting amped up with anxiety or excitement . . . you are draining yourself of your most important resource: energy."

Of course, we lead rich lives full of many experiences that generate emotions. What, then, is so special about the link between digital technology use and emotions? In the mid-1990s, David Mick and Susan Fournier, marketing professors at the University of Wisconsin and Harvard, respectively, embarked on an ambitious study in which they followed twenty-nine households whose members purchased new technological products like computers, televisions, and portable CD players (it was the '90s, after all!). Each time someone from one of the families bought a new technology, the research team interviewed them within twenty-four hours of the purchase to gain their initial reactions. Then they followed up with more interviews six weeks later, and then finally six months after the initial purchase. The researchers expected to learn how people made sense of new technologies and how they decided whether they liked or disliked their purchases over time. But they were unprepared for how much these technologies exacted "a troubling emotional toll" on the study participants.

Mick and Fournier's analysis suggested that the main drivers of such strong emotional reactions were the array of paradoxes people experienced when using the new technologies. For example, participants described how their new purchases inspired in them feelings of confidence and intelligence because they were using advanced technologies, while simultaneously leading them to feel ignorant or inept because they could not figure out many of the technology's features. Their use of the new technologies led them to feel more capable—that they had an enhanced ability to make the choices they wanted or find the information they needed—and simultaneously more constrained to work within the confines of the features the technologies provided. And they felt like their new products helped them to become much more efficient and reduced their effort in some areas, while also leading to greater investments of time and energy in others. It seemed that for every good thing the technologies brought into people's lives, they also created confusion and raised doubt and consternation so much so that the authors concluded that, on the whole, exposure to new technologies "wreaked emotional havoc, with feelings ranging from envy, foolishness, cautiousness, and frustration to fear, betrayal and defeat." It's important to note that this study was done before most people had internet in their homes, before most businesses had their records digitized, before there were two million apps in Apple's App Store (and before there was an App Store), before social media allowed us to see intimate details of others' personal lives and permitted others to see intimate details of ours, and before Google could connect us to the world's content and ChatGPT could summarize it for us. Needless to say, today we have far more opportunities for far more complex and paradoxical encounters with technology.

Our Janus-faced technologies pique our emotions so flagrantly because they make clear to us that we are not in control. The historian Merritt Roe Smith, who spent his career documenting how new technologies have affected the Western industrialized world, argues that all new innovations enter the world shrouded in a veil of technological determinism. Technological determinism is the belief that technological progress marches along according to its own logic and that we humans are just along for the ride. When we hear things like "AI is going to replace jobs," we're getting a dose of deterministic rhetoric. A few news stories or rumors from friends are not going to make us believe that our futures with technology are preordained. The problem is that technologically deterministic rhetoric doesn't find us just once, or even just every once in a while. Instead, we are confronted by the repeated insistence, year over year, technology by technology, that technological change is coming for us and that if we don't embrace it we are standing in the way of societal progress. When we begin to use our new technologies and experience the many paradoxes that Mick and Fournier describe, it only reinforces our preconception that we are not in charge. And that's a disquieting feeling.

So if we put two evidence-backed puzzle pieces together—1) that using new or complex technologies makes us feel out of control and 2) that lack of control is a strong predictor of heightened emotion—we would be foolish to ignore the role that digital technologies play in our emotional lives. I have identified five key emotions that are consequences and drivers of people's experience with the digital technologies they use each day. I think it's worth taking a look at these emotions to understand how and why our experience of them impacts our feelings of exhaustion.

FEAR

Sebastian is a top salesperson for a major electronics company. He's middle-aged, married, and a father of two. I was lucky enough to follow him around on the job for two weeks with my notepad and voice recorder. He was easy to "train" as an informant because he took quickly to the practice of speaking his thoughts out loud as he went about his daily tasks. At the end of the two-week period, I ran a linguistic analysis of my field notes and audio recordings. The program I used removed all nonsubstantive words—like prepositions and proper nouns—and revealed that the three words Sebastian said most frequently when talking about the technologies he used were "afraid," "scared," and "fear." Here are some of the things he said that included these words:

- "I'm always *afraid* there's some new technology I'm missing out on that would make my job easier. I'm always looking around for those."

- "We're always looking for ways to use AI better in our sales calls and I'm *afraid* it's either not gonna work and screw up things with the customer or it's gonna change how [the company] is gonna evaluate me."

- "Honestly, I get kind of *afraid* each time I hop on [the company's messaging platform] because there's just going to be so much shit I have to wade through and respond to. It just kind of makes me *scared*."

- "Then at this meeting they previewed like four new applications that they want us to use to track our expenses and stuff like that. I'm just like, each time there's some new technology to learn I get *scared* that it's gonna suck up all my time."

- "I hate LinkedIn, which is weird for someone in sales to say, but I hate it because I can't get away from it because what if I'm missing something important going on and I should have congratulated someone? That could cost me a relationship. I guess it's actually the same for all social media because I've kind of got that *fear* of missing out—FOMO, my kids call it—that maybe my sister-in-law's posting something on Facebook and then I won't see it and I'll look like an idiot 'cause I didn't know about it."

Sound familiar? Sebastian is a bright guy and certainly skilled enough to figure out each new device or application he's required to use. But the onslaught of new technologies and new sources of information coming at him every day provoked what he called "irrational fears." I don't think his fears are irrational at all. Countless other people I have interviewed across various age groups, industries, job types, and cultures voiced similar fears about their ability to continue learning new technologies and to keep up with all of the data and information those technologies send their way. As Jin, a pediatric surgeon at a major teaching hospital in Chicago, remarked, "You'd think as a surgeon I'd be most afraid of something bad happening in the OR. I'm actually most afraid that I'm going to screw up something major in EPIC [the hospital's electronic health record system] or miss out on some crucial information about my patients because there's just so many different systems we're trying out here to track different kinds of stats." Similarly, Kylie, a lawyer who works part-time so she can take care of her three school-age children, observed, "I'm always afraid I've missed something. Did the soccer field change? Did the coach send that information out in a text or was it in the GameChanger app? Is it an early dismissal from school? Was that on ParentSquare? Did the teacher email about it? I feel like I'm on pins and needles more at home than at work these days." Not surprisingly, Sebastian, Jin, and Kylie all scored high on the digital exhaustion scale. Although the fears associated with their technology use were by no means debilitating, they accumulated over time in ways that lead to Level 2 exhaustion.

There has been a significant amount of research into fears revolving around digital technology use. These studies sort into three categories. The first shows feelings of fear associated with the use of new technologies themselves. Needing to learn how to use new devices or applications seems to arouse in people the fear that they're not going to know which technologies to use or that they won't be able to learn to use them well enough to do their jobs, communicate with the right people, or get the best information they can. The second category shows that people fear the consequences of digital technologies.

These studies suggest that encountering new digital technologies like AI or social media—whether hearing about them or experiencing their capabilities—makes people fearful that their privacy will be compromised, that they'll lose their jobs, that their organization will restructure, or that their kids will access content that is too mature for them or become addicted to their smartphones or video games. The third category of studies shows that people fear missing out on important data or information. In these FOMO-centric studies, people recognize that they don't have enough time to keep up with all of the data and communications they can access through their digital devices, and they worry that they are missing out on important information. People appear to experience FOMO the most when they perceive they are missing information from a group with which they identify strongly—like missing a client update from members of their work team or missing photos discussed among a group of close friends. FOMO is a Sisyphean problem. The more people are exposed to information, the more likely they are to report that they fear they are missing out on important information. Studies show that FOMO is associated not only with mental and emotional exhaustion but also correlated with physical exhaustion. Adolescents who experience FOMO often stay up later and wake up more often at night to check their social media accounts, which of course makes them sleepy the next day.

ANXIETY

One of my favorite studies of all time tracked whether people bought jam after being exposed to in-store displays with varying quantities. The researchers found that people were more likely to buy jam from a local grocery store when the display showcased a small number of jams (six flavors) rather than a wide variety (twenty-four flavors). The study's conclusion, which has now been widely replicated in other contexts beyond jam-buying behavior, is that too much choice is de-motivating. Having too many options produces anxiety. Which one is best? How do I know? What if I make the wrong choice? When people experience anxiety they respond in a very rational way: They seek to eliminate their anxious feelings by simply not choosing. They don't buy any jam at all. Digital technologies create the same anxiety-inducing feelings. There are so many different technologies available to help us do the same tasks, and so much content to choose from, that it feels overwhelming to decide which to focus on.

Many nights, Cindy and her husband, Omar, sit down to watch a movie after putting their children to bed and doing the dishes. They turn on their smart TV, then the trouble begins. A typical night fol-lows a script that looks something like this:

8:30 Cindy and Omar flop onto the sofa in their living room.

8:32 Cindy opens the Netflix app.

8:37 Cindy scrolls through titles of TV shows and movies. She and Omar can't agree on one to watch.

8:43 Cindy convinces Omar to watch a preview for a British period drama. After watching the preview, Omar vetoes it. Cindy scrolls through more titles.

8:49 Cindy remembers an action movie they both said they wanted to watch. She searches for it on Netflix, but it is not currently available on the streaming service.

8:52 She switches to the Apple TV app and finds the movie for rent for $4.99.

8:54 Omar says he doesn't want to pay $4.99 for the movie and tells Cindy to check if it is free on Amazon Prime.

8:57 Cindy opens the Prime Video app and searches for the movie. It's also $4.99 there. Cindy tells Omar that if they're going to pay for a movie, she'd rather watch a different one. She describes it to Omar. She searches for it on the app and finds it is not available to rent, only to buy for $19.99.

9:04 Omar tells her to search Disney+ because he thinks it might be a Disney title and will be free there. Cindy throws the remote to Omar and tells him to do it.

9:07 Omar opens the Disney+ app but cannot find the movie there.

9:10 Omar scrolls through options on Disney+ and finds a TV show he suggests watching. Cindy asks to watch the preview. She says it looks good and they should watch it.

9:15 Omar notices that the show is fifty minutes long. Omar says he thinks it's too late to watch a fifty-minute show and tosses the remote to Cindy to find something else that is shorter. Maybe twenty-five minutes?

9:18 Cindy scrolls through several options then turns the TV off. "This is so annoying," she says. "I can't decide on anything. Let's just go to bed."

9:25 Cindy and Omar turn off the light and head to bed.

All that work for no movie! After switching among four video streaming apps over fifty minutes and seeing more than two hundred different movie options, Cindy and Omar decided to stop trying to decide. As Cindy observed when reflecting on her after-dinner movie-watching routine, "Lately I'm not even into watching a movie because the thought of having to pick one makes me anxious." Watching a movie to relax is supposed to be, well, relaxing, not anxiety producing. But the endless options that our digital technologies present to us make it difficult for us to decide what to pay attention to. And the realization that every choice negates other options that *could* have been better produces even more anxiety. The approach taken by Netflix and so many other digital tools that we use is to put more jam in front of us at just the moment when we're having a hard time choosing among too many jams.

Of course, it's not just the effortless and immediate availability of movies or songs or books or recipes or TikTok videos that brings anxiety, it's also the availability of much more boring things like news articles, data about our customers, market research analyses, or information about the uptime of machines in the plant. The realization that we have access to so much more data and information than we could ever sort through creates feelings of anxiety over how we choose. Health researchers have documented the rise of the "cyberchondriac," which was recently defined in the *Oxford English Dictionary* as "a person who (obsessively) researches health information on the internet." A cyberchondriac cannot be sated by the vast quantities of information about their potential condition and experiences anxiety due to the belief that they are missing out on better information somewhere else. Librarians regularly discuss how to help patrons deal with information anxiety, which Ashley Eklof, the head librarian at BiblioTech in San Antonio, Texas—the only public bookless library in America—says is "brought on by the desire to absorb as much information as possible, feeling overwhelmed by the amount of information being filtered to you." Our digital tools give us so many options to choose in every domain of our lives, which creates too many opportunities for us to worry about whether we are making the right choice.

Any discussion of digital technologies and anxieties inevitably turns toward the role of social media in our lives. There are many individual studies linking social media use—especially among adolescents and young adults—to anxiety, depression, and feelings of loneliness. Several recent meta-analyses have confirmed a robust relationship between increased social media use and increased anxiety. To Sierra, a woman in her early twenties who works as a technical sales representative for a software company, this correlation between social media use and anxiety makes sense. As she told me when discussing her use of Instagram, Snapchat, TikTok, and Venmo, "They're anxiety producing. All of them. Everyone is looking better than you and doing more fun things than you and you just get this feeling inside that you're not doing good enough. Honestly, it makes me anxious just talking about it." Gretchen, a seventy-year-old recent retiree and avid golfer, says, "I'm embarrassed to admit that social media kind of make me anxious. I go on Facebook or Nextdoor and the things people are doing or complaining about just really make me feel uncomfortable. I don't want to know all of these things. They get my pulse throbbing. But I can't look away. I'm like a moth to a flame."

No findings are without caveats, and the link between social media and anxiety is no exception. Research finds that people who are passive users (who read other people's posts and look at their pictures but rarely post themselves) tend to have more heightened feelings of anxiety on social media than people who are active posters. And a team from the University of Pittsburgh's School of Medicine looked at a nationally representative sample of adult social media users and found that only at extremely high levels of social media use did respondents show an increase in the odds of elevated anxiety and symptoms of depression.

Although these findings show that not everybody experiences anxiety while *on* social media, I would conjecture that most of us feel anxiety *about* social media. We're barraged with media reports about how executives at Meta knowingly covered up evidence that its Facebook and Instagram platforms are "toxic for teen girls." Former employees of social media companies, like Frances Haugen, who left Facebook and disclosed internal documents detailing how the company knew about the negative effects its platforms could have on the mental health of its users, and Tristan Harris, a former designer at Google who became a vocal critic of the way social media and other tech companies exploit users, have become household names. Documentaries like *The Social Dilemma*, about social media's addictive qualities, and *The Great Hack*, about the Facebook–Cambridge Analytica scandal where personal data was used to influence voter behavior, fill our screens. As Oliver, a geologist for a petrochemical company, put it, "You can't escape people talking about social media and how it's affecting the world. I mean I don't even really use social media and I'm anxious. All this talk about it just makes me worried and exhausted."

As I write this, there is so much anxiety-producing rhetoric about the effects of AI that is exhausting us. As Johan, CFO for a consumer goods company, recently old me, "I have no real idea how AI is going to change our business or whether it's going to put people out of work. But I worry about AI every day because you can't look left or right without hearing about AI and how it's going to change my business and put people out of work. I mean, all this talk about AI makes me anxious. Seriously!" And who could blame him? A study published in 2020 that investigated media coverage of AI showed that, despite a long history of coverage of AI in the mainstream media, articles discussing the positive and negative consequences of this emerging technology only became "intensive" beginning in 2015. Not coincidentally, the results of a Gallup survey conducted at repeated intervals between 2018 and 2022 showed that the share of people who say they are worried that AI will make their jobs obsolete is growing swiftly, which of course has led journalists to begin writing about "FOBO" (Fear of Becoming Obsolete). Although no systematic study has yet been conducted since the fall of 2023 when OpenAI released ChatGPT, I think it's safe to assume that what counted as "intensive" media coverage in the years prior to the 2020 study would look subdued by today's standards and that reported feelings of anxiety around AI are growing. The media love to create fervor around new and unsettled technologies. And few new technologies are as unsettled as AI. As famed astrophysicist Stephen Hawking once commented, "The rise of powerful AI will either be the best or the worst thing ever to happen to humanity. We do not yet know which."

GUILT

Jim, Steph, and Kelly are three siblings who grew up in a tight-knit family in the Southwestern US. Jim's the oldest, and Steph and Kelly are sisters, born less than a year apart. As the only boy and the oldest, separated from two close-in-age sisters by several years, Jim always felt that his sisters had a bond with each other that he didn't share. Nevertheless, the three always considered themselves close growing up. As they each graduated from high school and left home for college, work, and to start families in different states, they grew apart. Well, at least each one of them thought they were growing apart from the other two. When I spoke with Jim, he explained, "I see my sisters, Steph and Kelly, posting on Facebook a lot. One of them will post some picture of their kids or something and then the other one will respond with some joke that I don't understand. On the one hand it makes me happy to see that they're still so close, but then I feel guilty that I'm not putting as much into maintaining the relationship as they are. I know I should do better." But when I spoke with Kelly, she thought Jim and Steph were the ones with the closest relationship: "We've got this sibling text thread and I'm just not a great texter. Jim and Steph are always texting each other back and forth and making good little comments. They ask me things and I just take so long to respond. I feel guilty that they're taking the time to make sure they are in each other's lives, and I just don't do it as much."

The three siblings' feelings of guilt about their failure to maintain relationships were especially pronounced because they could see, or at least infer, a strong relationship between the other two siblings in their patterns of response on social media and text. A recent study led by Annabell Halfmann at the University of Mannheim showed experimental evidence that the frequency with which people use digital tools such as instant messengers is associated with more pronounced feelings of guilt, which could stem from the fact that they feel guilty about being on digital tools when they could or should be doing something else or that they feel guilty that they are not communicating and interacting in ways that correspond with socially accepted practices on the platform—for example, not responding quickly enough to messages or indicating that they like someone's post or photo.

Digital technology users often experience guilt that they are not doing as much as others. A common experience described by my interviewees was seeing coworkers who worked later or seemed to spend more hours on a project than they did and feeling guilty that they were not contributing sufficiently. As Fede, a professor at a New England university, recounted to me, "I've been having this experience more frequently lately where I'm working on a project like a grant application and we're using Google Docs so we can all collaborate on it. I go on there and I see my collaborators have been writing a ton and making all these changes. I feel guilty that I'm not contributing enough to the project. This seems to happen a lot lately. I'm constantly full of guilt on all my projects." An in-depth study of users of three immersive video game platforms found that feelings of guilt were very common among adult players who could see that other players had contributed in more significant ways to the game's quest, that their own actions had led to the death of another player, or that they were not contributing as much as others in their in-game family to making money to buy supplies, weapons, and special powers. Across all domains of life, digital technologies expose us to the behaviors and actions of others in unprecedented ways.

In my own research I've found that even though, empirically, we infer that other people are doing *less or worse* than us just as often as we infer that they are doing *more or better*, we tend to focus on what we lack or where we are deficient rather than focusing on the shortcomings of others. Our feelings of guilt are often more palpable and more memorable than our feelings of annoyance at others for shirking responsibilities. As Queen Gertrude remarks of Ophelia's erratic behavior in Shakespeare's *Hamlet*, "So full of artless jealousy is guilt, / It spills itself in fearing to be spilt." Our focus on personal deficiencies over the failings of others not only highlights our feelings of inadequacy, it also inadvertently exposes the guilt associated with them. This phenomenon serves as a compelling explanation for why we might feel guilt more intensely and memorably than irritation toward others.

ANGER

Anger was the most reflexive feeling I uncovered in interviews. People felt angry because they recognized that using digital technologies often made them feel fear, anxiety, and guilt. Their anger came because they didn't like these feelings, but they believed that the demands to use digital tools in their work and personal lives meant that they couldn't avoid them. As Rachel, a financial analyst for a large life insurance firm, recounted to me when discussing her digital exhaustion, "Sometimes I'm just angry at the way we have to live these days.

Our parents didn't have to balance so many texts and phone calls and emails and feeling like you're always missing out or you're guilty that you didn't help out enough with your friend's baby shower. I just get mad that this digital world makes me anxious and that my kids are anxious and that my team at work is anxious." Teddy, a creative designer for a real estate software company, shared similar feelings of anger when talking about his digital exhaustion: "I'd mostly say that I'm angry at what technologies are doing to us. Like I'm angry at how they make us all feel like we're not good enough and we're not doing enough. A lot of times when I think about all the technology in our lives I just get angry, and that wears me out."

In their timely book, *Bored, Lonely, Angry, Stupid: Changing Feelings About Technology from the Telegraph to Twitter*, Luke Fernandez and Susan Matt trace how norms about feelings have shifted alongside technological change. Through in-depth interviews and archival research, the authors demonstrate that our digital world stimulates and affirms a range of negative emotions. Users of new digital technologies find themselves angrier, on the whole, than users of analog media and consumers of mass media. They attribute a big part of these growing feelings of anger to the range of negative emotions that people experience when using, thinking about, and grappling with digital technologies. Fernandez and Matt chronicle how, throughout history, new technologies have consistently fueled hopes of reducing solitude and boredom yet often fail to fully deliver on these promises. Initially, new communication tools like the television brought communities together, creating shared experiences that countered loneliness. However, as these technologies became more commonplace, their ability to isolate individuals grew. Televisions were initially an excuse for families and neighbors to watch shows together, but it didn't take long for each person in a house to retreat to separate rooms to watch different TVs. This pattern has echoed into the digital era with technologies like social media and smartphones. But whereas television allowed us access to certain content at set intervals in particular locations, smartphones allow us access to whatever content we want, whenever and wherever we want it. Thus, while today's digital technologies offer temporary relief from boredom just like TVs did, they paradoxically lower our tolerance for it, making us less equipped to handle idle moments. People now get angry if they lose internet access and are bored. As Fernandez and Matt describe, it has become common for people to reflect on how digital technologies have changed the way they experience all manner of emotions and how they've grown increasingly angry at this revelation.

Anger is one of our most powerful emotions. It's also one of the most exhausting. When we're angry, our body launches into that familiar fight-or-flight response, increasing heart rate, blood pressure, and the release of stress hormones like adrenaline and cortisol. Yet studies have shown that anger appears to be somewhat unique among emotions in its effects on our fight-or-flight response. It seems to delay the return to a normal physiological state, maintaining high levels of stress hormones that contribute to fatigue. Anger doesn't just tire our bodies—it also tires our minds. The cognitive load of anger is significant; maintaining focus on the source of irritation and planning responses eat up cognitive resources. Anger is also one of the hardest emotions to control. As Emma Seppälä says, "Self-control actually exhausts us, it's a limited resource like gasoline or the charge on your cellphone. The more you use it, the less you have it. Researchers have found that it literally depletes your blood sugar. Ever wondered why you are more likely to binge on ice cream at night? Self-control literally gets depleted as the day goes on." This combination of physical and cognitive exhaustion, and the ever more exhausting attempts to control our anger, are why people like Rachel and Teddy are so worn out by the anger they feel when they recognize how digital tools provoke so many negative emotions.

EXCITEMENT

Technology doesn't always elicit solely negative emotions. New technologies are cool. They're fun to use. Whether you're exploring how a new photo app can change the look of your bedroom or asking Chat-GPT to write couplets about sandwiches, it's easy to get excited about a new tool's capabilities. If you or your kid has powered through seven straight hours of play on a new video game, or if you've rushed to set up your new iPad or computer, you know how exciting it can be to use new technology.

But even our experience of excitement is exhausting. Those same stress hormones that kick off our fight-or-flight response to fear and anger also increase production when we feel excited. When we're excited about something, our pulse quickens, cortisol increases, and— you guessed it—those physiological reactions prove exhausting. Excitement also compels us to do more on our digital tools—to search for more information, download another app, post another photo, text another person, watch another video—which as we know are other pathways to exhaustion.

Earlier in this chapter, I highlighted several studies that show a clear link between exhaustion and various emotions people experience when using digital tools. There are few good studies that show the positive ways that digital technologies exhaust us. When studies show that Facebook use among college students increases social capital, that using a knowledge management technology at work improves knowledge transfer, that Instagram connects old friends, or that

ChatGPT rekindles someone's interest in short stories, they demonstrate what is gained by our digital technology use but not what is lost. This may be because most studies of the positive benefits of digital technologies focus on their practical benefits, which are presumed to be more connections with people, faster knowledge integration, and so on. But my interviews and observations of digital technology users across all areas of work and leisure show that practical benefits arrive alongside emotional ones.

Take Joni, a nurse at a Midwestern hospital. She and her nursing colleagues had just gotten access to a new data portal that provided detailed information on patients who were going to be transferred from other hospitals. As Joni recalled after one week of using the new system, "My heart is literally pounding every time I open it. I just can't believe how much information we have about the patients now before they arrive. It is incredible. I am so excited. This is such a game changer." Or consider the reaction of Micah, a young software engineer whose team just started using Slack, "I'm so pumped we have Slack now. Like I seriously have to control my excitement. Tracking what's going on with the team just got so much easier. We're going to function so much better. This is just really cool." Or even Rick, a father of a newborn girl, when he got this first iPhone: "This thing is sweet. I love it. I kind of get giddy whenever I pick it up. My wife says I get more excited about the phone than our daughter. She's joking, though. Really. But I get so pumped when I can just snap pictures of her [his daughter] whenever she's doing something cute." Across all of these examples and countless more, I've seen people bursting with joy at doing something with new digital tools that they couldn't do before. That's exciting!

Access to new data, new capabilities, and new "cool" technologies excites us. And it should. The reason we use digital tools is because they offer capabilities that make our lives better, easier, and more fulfilling.

We should celebrate that. But it's important to keep in mind that such excitement can take its toll.

It's not our digital technologies that exhaust us, but rather the ways we pay attention, make inferences, and experience emotions with and through them. Understanding how the exhaustion triad works to sap our energy while giving us immense capabilities is a crucial step to figuring out how we defeat our digital exhaustion. In the next section, I provide eight simple rules that will help us to counteract the insidious forces that work together to exhaust us.

PART II

Simple Rules
for Resilience

Stop Using Half Your Tools

Shireen is a marketing specialist at a start-up company that produces and installs solar panels. One foggy day in May, I asked her to make a count of all the digital computer programs and apps she used that day—at work and at home. Here's her list:

1. Microsoft Word
2. Adobe Illustrator
3. PowerPoint
4. Canva
5. Outlook
6. Gmail (via web browser)
7. Chrome
8. HubSpot
9. Salesforce
10. Zoom
11. Google (search)
12. Microsoft Teams
13. Jira
14. Twitch
15. Instagram
16. TikTok
17. (Apple) Mail
18. ChatGPT
19. SharePoint
20. WhatsApp
21. iMessage
22. Dropbox
23. Amazon app
24. Chase Mobile app
25. Twitter
26. LinkedIn
27. Spotify

28. Evernote	31. Color Switch	34. Hulu
29. Waze	32. Timehop	35. Siri
30. Nike Run Club	33. Netflix	36. Alexa

She uses these programs and apps across several physical devices, including her laptop, smartphone, Apple Watch, Echo Show, smart TV, and the smartphone interface in her car. After compiling the list, Shireen commented, "Oh, God. That exhausts me just looking at it."

If you and I made lists like Shireen's, they wouldn't look too different. I know this because I've asked more than two hundred people to make such lists over the years. In the 2000s, the average number of digital tools that people reported was eight. In the 2010s that number jumped to twenty-five, and in the 2020s that number rose to thirty-four. In the mid-2000s, one of the "digital technologies" that just about everyone I interviewed listed was a cell phone. Most people discussed calling others on their cell phone, and a few people talked about texting with it. But there were no apps, no App Store, and people were not accessing content through their phones. Some people in those early interviews also mentioned laptops as one of their digital technologies, and they made a distinction between the kinds of activities they would do on their desktop versus those on their laptop. Twenty years ago, those distinctions made sense. Cell phones were not smart, laptops didn't have nearly the power or storage capacity of desktop computers, and there was limited access to Wi-Fi even if your phone could have been smart and your laptop powerful enough to make use of it in meaningful ways. Today, no one I ask to list their digital technologies even mentions the devices they carry. It's just assumed that you'll be able to access the applications or programs you need at any time in any location. Most people think of devices as access points to digital technologies rather than digital technologies themselves.

As we discussed in chapter 1, using so many different applications and devices creates a significant cognitive load, and frequently switching between apps is a key driver of exhaustion. One study published in 2010 found that increases in the number of digital technologies that knowledge workers used increased their productivity only up to a point, after which using more digital technologies lowered productivity. These findings make intuitive sense. If a digital tool provides us with a critical capability, we're likely to be more productive. But as humans we can juggle the cognitive demands of only so many tools. After we reach a certain point of capability enhancement, we simply cannot absorb more. The marginal benefits of each new capability are lost on us. And worse, the demands placed upon us to learn, use, and switch among a larger portfolio of digital tools begin to exhaust us. That point at which our digital tools switch from providing enhanced capabilities

More Digital Tools Enhance Our Productivity– But Only up to a Point

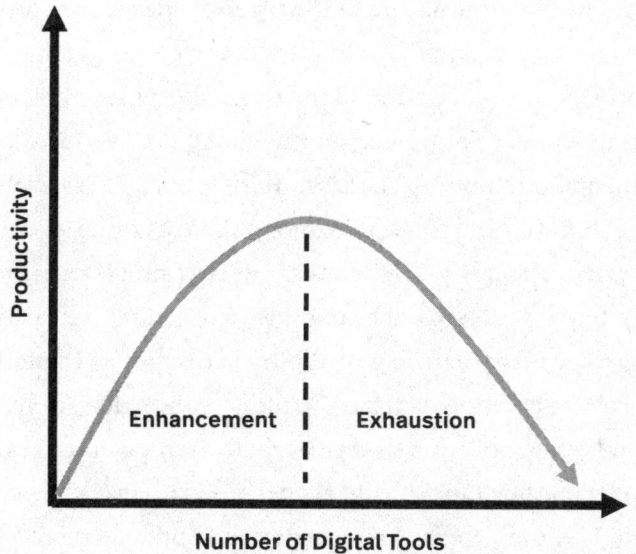

Enhancement | Exhaustion

Number of Digital Tools

to being sources of exhaustion is different for each of us and is related to the degree to which our work depends on technology. The more that we rely upon specific digital tools to do our work, the more feelings of digital exhaustion will negatively impact our productivity.

In 2010, the digital technologies we used were relatively stable. You bought a license for a particular version of Microsoft Excel or Adobe Photoshop, for example, and with the exception of minor bug updates pushed to your computer periodically, the software stayed the same. Once you learned to use it, you knew how to use it. But in today's world of rapid technological advancement in which almost all software applications that we use are accessed via the web and are updated constantly, the features embedded in technologies are updated and changed frequently. That means we are always learning and relearning how to use our digital tools. In 2020, a research team conducted a study to uncover whether the learning we must do in the face of constant feature updates leads to exhaustion. Their survey of 489 Facebook users showed that it indeed does. The findings provide evidence that it's not just learning how to use new digital tools that exhausts us. The dynamic nature of the technologies we use forces us to continuously cope with change in ways that wear us out.

Most of us use entirely too many technologies, each of which has too many features that are constantly changing. We just don't have enough mental or emotional bandwidth to learn and relearn so many different interfaces and to switch among them as often as we do with any semblance of alacrity. This leads to our first simple rule: Stop using half your tools. Reducing the amount of learning, relearning, and switching that you have to do will allow you to enjoy the benefits of the technological capabilities you have while keeping exhaustion at bay.

When I tell people this first rule, they often say it makes sense to them. But then they quickly add that there is no way they could just stop using half their tools. An experiment I conducted with Rebecca

Hinds of Asana, Federico Torreti of Amazon Web Services, and Bob Sutton of Stanford University illustrates just how hard giving up your tools can be. We asked fifty-eight employees from Asana and Amazon to list all of the digital collaboration technologies they used at least once a week for interacting with colleagues. We also asked them to rate how well each technology helped them to achieve their work goals and how hard it was to use. Of course, we also asked them the digital exhaustion question. We then randomly split them into two groups. We asked people in the first group to stop using half their tools for two weeks. We told the people in the second group to choose how many technologies to eliminate. Both groups chose which tools to eliminate and submitted a list to us. Throughout the study they kept daily diaries of which tools they used and explanations of why they might have used any tools that were on their "do not use" list.

The experiment was a disaster. Pretty much no one in either group stuck to their lists. Even the people who committed to giving up half their tools ended up using most of them anyway. The participants told us how hard it was to not use digital tools they were accustomed to. Their diary entries revealed that their bosses demanded that they use some technologies that they had hoped to give up. They also described how many coworkers simply would not respect the fact that they chose to give up certain collaboration tools like Slack when it was the team's norm to use those tools to communicate or provide project updates. All of that was bad. But the worst part was that at the end of the study we asked participants to rate their levels of digital exhaustion again. The average level actually *increased* for most people over the two-week period, and the rate of increase was higher for the people in the group who committed to giving up half their tools than it was in the group who just got to pick which tools they would give up.

When we probed deeper into these findings about why their exhaustion increased, one answer became obvious. The experiment revealed to participants just how little control they felt they had over which digital tools they used. As we learned in chapter 3, feeling out of control about one's own ability to decide which digital technologies to use and how to use them is a key driver of exhaustion. As they tried to give up various digital technologies, our participants found that using so many tools was indeed exhausting. As one participant noted, "[The study] made me much more aware of the different tools I was using and the impact that was having on my productivity and sense of focus. There is a subtle cost with switching tools, and that additional friction hurts my focus and productivity." Recognizing that they wanted to use fewer tools but then finding that they were unable to fully control which tools they used led these workers to feel even more exhausted and, as one participant noted, "out of control."

The benefit of eliminating a significant portion of one's digital technologies has good scientific backing. In their recent book *The Friction Project*, Bob Sutton and his longtime collaborator Huggy Rao describe numerous studies that show how subtracting rather than adding rules, policies, processes, jargon, or even technologies makes things better. But as Sutton and Rao highlight, when given the choice we tend to add rather than subtract. They call this tendency "addition sickness." As a couple of examples of this problem, they recount how only 11 percent of faculty recommended subtracting processes in response to a university president's call for ideas on how to improve the institution, and how most participants tasked with modifying LEGO structures to withstand the weight of a masonry brick added more LEGO bricks (even though they were charged for each additional brick) when the best solution was to remove one. Science was on our side, but we hadn't given the participants in our study enough guidance on how to subtract their tools.

Since that initial experiment, I've followed up with more than fifty workers across a variety of roles and industries who were willing to try to stop using half their digital tools. But this time, I gave them explicit instructions on how to subtract successfully. After six months, 86 percent of people were still using just half of their technologies, and digital exhaustion scores were down by an average of 40 percent. Below, I outline what they did to subtract half their digital tools and make the cuts stick. To make it more concrete, let's follow Shireen as an example.

STEP 1: DETERMINE WHICH TOOLS YOU USE AT WORK AND AT HOME

The first step is to figure out which digital technologies you use for work, which you use exclusively for leisure or to coordinate your personal life (what I'll call "home" for shorthand), and which you use for both. In Shireen's list, those used exclusively at work are highlighted in bold, those used exclusively outside of work are in plain text, and those used for both work and home are in italics. Most people are surprised to see that their list looks like Shireen's. We think that the workplace puts many more demands on which digital tools we must use than we put on ourselves outside of work, but that's not typically true. Most of us choose to use a dizzying number of applications and platforms at home. Although we might convince ourselves that those outside-of-work tools are necessary for us to live fulfilling lives, most of them are not. The good news about a list that looks like this is that we have much more control over eliminating technologies we use at home than we do for technologies we use at work. Shireen's list shows that twenty-one out of thirty-six—more than half—of the digital technologies she uses are ones that she can decide to stop using without the threat of being reprimanded by a boss or fired.

For roughly half of the people I've worked with on this exercise, just completing this first step was enough to lower their exhaustion scores. Why? It goes back to those feelings of control that we discussed earlier. When you recognize that roughly half of the technologies you use daily are ones you've chosen and can decide to stop using if you want, you feel more in control. You know you *could* subtract them if you wanted to and, incredibly, that's enough to reduce your feelings of exhaustion. Although this reduction is most pronounced right when people realize that they have control, its effects linger so long as you know that you have the power to make a change if you so choose. Of course, those reductions in exhaustion will be much more profound if you continue with the following steps.

1. *Microsoft Word*

2. Adobe Illustrator

3. PowerPoint

4. Canva

5. Outlook

6. Gmail (via web browser)

7. *Chrome*

8. HubSpot

9. Salesforce

10. Zoom

11. *Google (search)*

12. Microsoft Teams

13. Jira

14. Twitch

15. Instagram

16. TikTok

17. (Apple) Mail

18. *ChatGPT*

19. SharePoint

20. WhatsApp

21. iMessage

22. Dropbox

23. Amazon app

24. Chase Mobile app

25. Twitter

26. LinkedIn

27. Spotify

28. Evernote

29. Waze

30. Nike Run Club

31. Color Switch

32. Timehop

33. Netflix

34. Hulu

35. Siri

36. Alexa

STEP 2: IDENTIFY NON-SUBSTITUTABILITY, INSTRUMENTALITY, AND NETWORK LOCK-IN

For step 2 we need to focus on the characteristic of each tool. It's best to start with the home list because, as we discussed, it's easiest to make changes here and we typically have just as many if not more home tools than work tools, so it's easy to make significant progress. The first characteristic to examine is substitutability, or whether any two digital tools on the list can be substitutes for each other. For example, on Shireen's list, Gmail and Apple Mail are substitutes because they both give her access to non-work emails. Shireen used Apple Mail as a client to read her Gmail on her iPhone, but she could just use the Gmail app, which she does on her computer anyway. Siri and Alexa are substitutes too, because Shireen basically uses these AI-powered assistants for fact-checking and to listen to music. Instagram, TikTok, and Timehop are all apps that allow her to share photos and videos with friends. Twitter and LinkedIn are also substitutes. Even though they reach slightly different audiences, going through the work of step 2 helped Shireen realize that almost everyone she follows on Twitter is also on LinkedIn. Using both platforms offers her relatively few unique advantages over just using one of them.

The second characteristic to examine is instrumentality. Some digital tools are instrumental in helping us do things we want to do. We *need* to use them because we have to do a certain thing or there are simply no other options. Some tools clearly fall into this category. If Shireen wants to buy something from Amazon or check the balance on her Chase credit cards, she needs to use those apps (or go directly to the vendor's website, which is less convenient and no less cognitively demanding than just using the app); there isn't another choice. But she doesn't need to play Color Switch, and she doesn't use any of the advanced functionality in Evernote beyond just taking notes, which she could do in MS Word. And she realizes that although she spends a lot of time surfing Twitch and Hulu to stream shows and movies, she almost never decides to watch anything other than options available to her in Netflix. All of these digital tools provide capabilities for sure, but they are not instrumental capabilities that Shireen utilizes. So those are easy candidates for subtraction.

The third characteristic is to look for network lock-in. As a network grows in size, its increased value makes users more dependent on it for their communication needs. And as more people become dependent on it, the value of the network grows. This phenomenon is called Metcalfe's law. Although not all of the digital technologies we use directly connect people in social networks, many of them lock us into them, as opposed to alternatives, because of networks. Take Word, for example. You might not want to use it. But because the other people in your network—at work or at home—use it and send you Word files, you can't not use it. You're locked in. If all your friends outside the US are on WhatsApp for mobile messaging, you have to use WhatsApp if you want to text with them. And, of course, navigation apps like Waze are superior to competitors because they have so many users voluntarily or involuntarily contributing data to the platform about how long it takes to drive from point A to point B or whether there is an accident on a certain part of the highway. The functionality of the technology makes it difficult to move on to another platform with fewer users/contributors. All of these examples show how we can become locked into the use of certain technologies because we want or need the interactions that they foster. Giving them up doesn't mean just losing the capabilities the technologies provide; it also means losing the people and insights those technologies make available to us. When Shireen looks at her list, she sees that most of her digital tools have her locked into network dynamics. However, she realizes that the native exercise app on her Apple Watch is a substitute for most of the functionality of Nike Run Club, except for the community data the app provides. But because she doesn't really ever use any of that data, this is an application to subtract. Further, she has so few interactions with friends on WhatsApp that she often struggles to remember how to use basic features in it. If she downloads an integration program she can easily check and send WhatsApp messages through iMessage, eliminating the need to check that app too.

To complete step 2, you should look at the work tools on your list to identify non-substitutability, instrumentality, and network lock-in. Then, ask yourself which digital technologies that do not meet those three criteria you can subtract from your tech stack. Shireen determined that just like she did with Apple Mail and Gmail, she could stop using Outlook as an email client and just use the Gmail web interface instead since her company uses a Google email solution. She also used Canva for graphic design work for early stages of projects because it was what she is accustomed to using from her last employer. But her company had access to the more powerful Adobe Illustrator, and she was eventually required to submit all her projects in that format anyway, so she decided she could just stop using Canva. Although the degrees of freedom she had with her work tools were fewer than with her home tools, when we look at Shireen's list after step 2 (tools eliminated are crossed out), we can see that she has been able to stop using thirteen of her thirty-six tools. Not bad.

1. *Microsoft Word*	5. ~~Outlook~~	9. **Salesforce**
2. **Adobe Illustrator**	6. Gmail (via web browser)	10. **Zoom**
3. **PowerPoint**	7. *Chrome*	11. *Google (search)*
4. ~~Canva~~	8. **HubSpot**	12. **Microsoft Teams**

13. Jira	21. iMessage	29. Waze
~~14. Twitch~~	**22. Dropbox**	~~30. Nike Run Club~~
15. Instagram	23. Amazon app	~~31. Color Switch~~
~~16. TikTok~~	24. Chase Mobile app	~~32. Timehop~~
~~17. (Apple) Mail~~	~~25. Twitter~~	33. Netflix
18. ChatGPT	26. LinkedIn	~~34. Hulu~~
19. SharePoint	27. Spotify	35. Siri
~~20. WhatsApp~~	~~28. Evernote~~	~~36. Alexa~~

STEP 3: FOCUS WHERE YOU HAVE INFLUENCE TO MAKE A CHANGE

When focusing on the remaining tools at work, it may seem that you don't have much ability to make changes. If your company, like Shireen's, has decided to use Salesforce as a customer relationship management tool and your work intersects with customers, you're probably going to have to use Salesforce. Similarly, if your team coordinates through Microsoft Teams chats and you're the one person who's opted out, you're not going to be a very good team member. In these cases, your work tools are non-substitutable, instrumental, and fortified by network lock-in. To be able to stop using these kinds of digital tools, you'd need other people to also agree to stop using them. Shireen didn't have the authority to make such changes, and many of us likely don't either.

But there are plenty of digital tools that are still in use in the workplace because people have developed habits around using them, not because they meet any of the criteria outlined in step 2. There's a long line of research on digital technology use in companies that shows that after making reasoned and rational decisions about whether to *start* using a new technology, people often *continue* using the technology because, as so many people have told me, "That's the tool that everyone uses." In fact, a classic study by Janet Fulk from the University Southern California showed that the more people identified with their work group, the more likely they were to continue using a digital messaging system just because they thought everyone else in the work group liked it. It didn't matter whether everyone else in the group actually liked the technology and wanted to use it. Social influence and habit are strong predictors of people's propensity to continue to use technologies without critical thought.

When you're not in a position of authority, the key is to find those technologies that do not meet those three criteria and create opportunities for change among your work colleagues. Shireen was able to eliminate four more digital tools from her list by taking this step. As one example, she realized that for reasons she did not understand, her company provided two options for saving files in shared locations. One was on the company's SharePoint server and the other was via Dropbox. Shireen used both but didn't have a good rationale for why. When she asked her other teammates, she found that most people didn't like SharePoint and preferred Dropbox. She asked her manager if there would be any harm in having the team use Dropbox rather than SharePoint. Her boss said he also was in the dark as to why they had two options and said that the switch was fine by him. So the team moved their files from SharePoint to Dropbox, and Shireen crossed one more digital technology off her list. In reflecting on her work tools, Shireen realized that she did not have any substantive work to do in Salesforce or Jira. She would log in to track some things that were documented in those systems, but her job did not require her to use them frequently. Because she didn't use them frequently, she felt considerable strain trying to remember their functionality each time she logged in. Searching for what she needed was harder than she felt it should be. She talked to her boss about this too, and he told her that he always sent messages to the team if there were things the team needed to know, so she didn't have to use these. Two more off the list.

Shireen's biggest gamble came when she thought about the way she used Microsoft Teams and Zoom. Her team messaged on Teams all the time, so she needed to use it for that function. But the team also used the video conferencing features often. She felt comfortable doing video meetings on Teams. She could blur her background and share her screen easily, and she knew how to adjust the volume on her microphone or speakers if there was a problem. She didn't have the same facility with Zoom. But two of the consultants she worked with on marketing collateral always sent Zoom invites for videoconferences. Shireen decided that because she hired those consultants, she could likely choose what platform they used for meetings. She kindly informed both consultants that in the future they would be conducting their videoconferences on Teams. She told them that she'd be happy to send the invites, but to her surprise, both responded and told her that they already used Teams with other clients so they'd be happy to switch to Teams meetings. As Shireen told me, "I am so glad I got over my nervousness about it and just told them I needed to switch to Teams. Who knows why they chose to use Zoom with me, but that couldn't have worked out better. I'd still be stuck on Zoom if I hadn't spoken up."

After completing this third step, Shireen had reduced the number of digital technologies she used on a daily basis from thirty-six to nineteen. OK, that's not exactly half. But as you've already surmised, getting to half was not really the goal. The point of this exercise is to make a substantial and meaningful reduction in the number of digital tools that we have to learn, relearn, and switch between. A reduction of 50 percent is a guideline. Twenty years ago, when people were averaging eight digital tools in their portfolio, eliminating half of your tools might have been too much. In ten years, it might be too little. The point is that it's easy to add and it's even easier to forget to subtract. As Leidy Klotz, author of the book *Subtract*, reminds us, we are wired to overlook subtraction, and we use addition as a substitute for thinking. We're often told that if we have a problem, it's because we haven't found the right solution yet or we don't have the right product. Don't think, just add. But when we're adding and not subtracting, our exhaustion grows. That's why we need to periodically cull tools from our tech stack.

1. *Microsoft Word*	**12. Microsoft Teams**	24. Chase Mobile app
2. Adobe Illustrator	**13. Jira**	~~25. Twitter~~
3. PowerPoint	~~14. Twitch~~	26. LinkedIn
~~4. Canva~~	15. Instagram	27. Spotify
~~5. Outlook~~	~~16. TikTok~~	~~28. Evernote~~
6. Gmail (via web browser)	~~17. (Apple) Mail~~	29. Waze
7. *Chrome*	18. *ChatGPT*	~~30. Nike Run Club~~
~~8. HubSpot~~	**~~19. SharePoint~~**	~~31. Color Switch~~
~~9. Salesforce~~	~~20. WhatsApp~~	~~32. Timehop~~
~~10. Zoom~~	21. iMessage	33. Netflix
11. *Google (search)*	**22. Dropbox**	~~34. Hulu~~
	23. Amazon app	35. Siri
		~~36. Alexa~~

STEP 4: FINALIZE YOUR LIST
WITH AN ENERGY AUDIT

The last step to finalize the list of digital technologies you're going to eliminate is to conduct an energy audit. It's an idea that Sarah Sarkis, a psychologist and performance coach, uses to help her clients figure out what drives and what saps their energy. "Energy, just like money, is finite," Saris argues. "You have credits, and you have debits. Any time you do something that benefits your mental or physical health, like sleeping or exercising, you get a credit. But any activities that are a detriment to that, like working late or skipping a meal, are debits." We can apply this idea of energy credits and debits to the digital tools on our list. Take all the tools that you've marked for subtraction and categorize each one as an "energy credit" or an "energy debit." Now make two columns, one for credits and one for debits. If using the tool *gives* you energy—it provides content you enjoy, helps you do some important task, makes you feel happy or excited, or connects you to people or content that stimulate you or warm your soul—put it in the credit column. It's an energy stimulant. But if you find that using a particular tool *takes* energy—you worry about using it well, can't remember its features, hate the aesthetics, or are bored or enraged by its content—put it in the debit column. It's an energy sapper. If the work you did in steps 1–3 placed it on your potential subtraction list and it also ended up in the energy debit column, that's even more reason to stop using it. But if you had it on your list for subtraction and it ended up in the energy credit column, you might rethink whether to eliminate it.

The only digital tool from Shireen's subtraction list that ended up in the energy credit column was TikTok. As she told me, "I know those videos are dumb, but I just like scrolling through them. I try really hard not to take it all so seriously and just enjoy it for purely stupid entertainment value. I mean usually when I'm done watching a few videos I'm just laughing and I feel good." If a technology in the credit column makes you feel the way Shireen feels about TikTok, it's probably a good candidate to retain. Remember our battery metaphor: We have to focus on stopping the forces that drain our energy and amplify those that add to it. Sometimes adding one more tool to our portfolio can be the right move as long as we add with purpose—not as a substitute for thinking.

Make a Match

When Claudia started her dream job as a paralegal at a prestigious law firm, she couldn't believe how inundated she was with communications from across the company. Messages came via email, Slack, text, the company's intranet, and DMs on the company's social media site. But what she quickly noticed was that almost no one contacted her via her office phone or by visiting her desk. "It was so weird," she said. "It's like everyone wanted me to respond to something, but no one actually wanted to talk to me—like, hear my voice."

After two months at the firm, the volume of communication became too intense, and Claudia found herself experiencing high levels of exhaustion. A major part of the problem was that questions that came to her via email, text, or Slack weren't always clear, so she had to have several back-and-forth emails, texts, or DMs with the sender to clarify. Her responses weren't always clear either, resulting in more back-and-forth. Exhausted and demotivated, Claudia tried a new tactic. When she received an ambiguous email, she picked up the phone and contacted the sender to get clarification. Two minutes on the phone was usually enough to bring clarity. By simply picking up the phone more often, Claudia reduced the number of messages she sent back and forth by more than 200 percent. "I feel so light now," Claudia said six months after making the switch. "Some people still think it's weird that I use the phone so much, but they don't actually mind and I have much better energy."

Claudia's solution to her exhaustion problem involved a simple but effective change in behavior: matching the capabilities of a particular technology with the ambiguity of the message she was trying to communicate. In the digital age, we tend to favor written, asynchronous communication channels in most cases. (Think of the last time your phone rang and you thought, "Why are they calling me?") But many types of messages and data are best dealt with synchronously and with the aid of nonverbal cues, rather than via text. Also, as the ongoing backlash against meetings and the society-wide discussions about remote work show, we often choose the wrong method of communication for the job, assuming that everything that is important should be done synchronously and in person. This is why our second rule to reduce digital exhaustion is to strategically match the technology you use with the complexity of the data you're dealing with and the coordination needs of the people working together. This matching should also take into account the symbolism certain digital technologies carry in your workplace, occupation, friend group, or family.

To understand the idea of matching, we first need to discuss a strange word: "affordances." Affordances refer to the actions that we believe we can take with a new technology. The idea is that any object has properties or features. A rock is solid, heavy, and round. A software program can compute standard mathematical functions, summing, averaging, creating amortization tables, and the like. Those properties or features are not perceived the same by everyone. A big animal might see a rock as a place to sit and rest. A small animal might see the rock as a safe structure under which to hide. Because of their physical differences, the two animals will perceive the same features of the rock differently. In that way, the same rock can afford one animal a place to sit and another a place to hide. For different animals, different activities are possible with the same rock. It turns out Facebook and your company's CRM system are not that different from rocks: The potential uses of a digital technology can be perceived very differently by two different people, who will thus use it very differently. An administrative assistant for a PR firm might perceive a spreadsheet software program like Excel as a tool to maintain and sort addresses and to perform mail merges, while a loan officer at a bank might see Excel as a tool to determine how large a mortgage his client can reasonably handle. Even though the software program is the same, it affords the assistant and the loan officer different capabilities because they each approach it with different goals and skills. Affordances are not a property of the tool per se but rather an outcome of the relationship between the person and the technology's capabilities.

The concept of affordance is core to Rule #2. If we're going to use our digital technologies in ways that keep the interactive forces of the exhaustion triad at bay, we need to match the affordances they present to us with some key features of our context. In the 1980s, Richard Daft and Robert Lengel came up with a scheme for thinking about matching. Although their "media richness theory" has been much debated and there has been inconclusive evidence supporting some of its original propositions, it provides a general framework that is quite useful. I've used it with great success with people in many walks of life.

The premise is simple. Digital technologies vary in the richness with which they can convey data and information. "Richness" refers to the number of cues they can convey. According to Daft and Lengel, the original technology for information dissemination and data processing was our body. When people are engaged in face-to-face interaction, they have many cues they can use to communicate and signal understanding, agreement, and buy-in or confusion, opposition, and dissent. The tone of our voice, the tilt of our heads, the furl of our brows, the pucker of our lips all accompany the words we utter as we convey and receive data. We can respond to the feedback we get as we share information and adjust our delivery or modulate our response in the moment. Because humans only had face-to-face interaction as a means for data transfer for millions of years, we have evolved in ways that make us uniquely attuned to process complex or ambiguous information in this way. In-person interaction affords discussion and convergence on meaning. For this reason, Daft and Lengel argued, face-to-face interaction was the richest communication medium. They then suggested that whatever technologies approximate face-to-face communication in their ability to allow real-time processing of multiple, complex cues would be the next richest media. The fewer cues a technology can convey or process in a meaningful way, the leaner the medium. Their typology (updated to reflect the technologies we typically use today) looks something like this.

Continuum of Media Richness

Lean — Email: Data (delayed) · Social: Chat, Data (real-time) · Phone: Data, Emotion · Video: Data, Emotion, Place · Face-to-Face: Data, Emotion, Place, Context — **Rich**

Other researchers argue about what counts as "richness" and what the right order of technologies might look like from lean to rich. I think those arguments obscure the larger point: We perceive that different digital technologies offer certain ways of sharing and processing information and make others difficult.

Sometimes the information that we're communicating or trying to make sense of is difficult to understand. Sometimes we're not sure how other people are making sense of it. And sometimes it's important to come to a joint understanding quickly because what we can do depends on what others think. But other times, the information or data we're working with are simple and self-explanatory, we don't need to make quick decisions, and we don't need people to agree with us in order to move forward. When we don't match the affordances of our digital technologies appropriately with the demands of the situation, we create a misalignment that exhausts us. In what follows we'll discuss how to make the right match and overcome the inertia that stops us from doing what seems so easy and intuitive.

CRYSTAL CLEAR OR CLEAR AS MUD?

I spent two months in Austria for work—a nine-hour time difference from my colleagues in Santa Barbara. I began emailing one of my colleagues, who we'll call Jeanette, at the end of the Austrian day for updates and to sort through some important personnel and budget matters. These were tricky subjects. On the personnel side, it wasn't clear what the best move forward was. We had to decide if we were going to let someone go and whether we were going to promote someone else. The best path forward in both cases was not crystal clear. We didn't see eye to eye on the next steps. On the budget side, Jeanette and I had differing interpretations of the numbers. I thought the numbers were bad; she thought they were good. We were looking at the same numbers, but we were orienting to them in different ways. Neither she nor I really understood that the other person's interpretations of the numbers differed from our own. I couldn't understand why she was making some suggestions about how to allocate our funds. From my perspective that the numbers were bad, her suggestions made no sense. And from Jeanette's perspective that the numbers were good, she couldn't understand why I kept making a big deal about certain things.

When I first started emailing Jeanette about these two issues, she responded right away. Right away meant by my next morning, because I'd email her at 6 p.m. in Austria, after I finished my other work in Vienna, which was 9 a.m. her time. But after a few days, her responses almost stopped. I grew frustrated. I started sending more frequent emails and texting her to ask what was going on. This interaction proved to be a major source of exhaustion for us because misunderstandings festered. We both became angry and frustrated that the other person was not understanding us, we used the minimal information available from those texts and emails to make determinations about the other's motives that were not very nice, and the constant emails and texts we sent back and forth to try to resolve the issue increased our need to frequently context switch out of the many other tasks we were each trying to accomplish so we could respond, yet again, to the other person. This particular interaction, which lasted more than two weeks, was extremely exhausting. The problem was that I was not matching the right technology with the nature of the issues we were dealing with. Neither was Jeanette.

The decisions we had to make around the personnel matters and the budget allocations were characterized by high degrees of equivocality. Equivocality occurs when information or data are clear but they hold different meanings to different people. Jeanette and I were both looking at the same data about our staff members' performance, but we saw different things. Where she saw improvement, I saw lack of commitment. And when we looked at the same budget proposals, she saw reasonable proposed expenditures and accounts in the black, while I saw requests for money that didn't seem to match the priorities of our department and account balances that were far too low to warrant big gambles in the coming year. In both cases the data were the same, but we saw different things. The data were equivocal.

Equivocality is not a problem in and of itself. Many situations in life are equivocal. And it's not just at work. Crew and Dalia are two parents I interviewed who discussed a fight they got into over text message while Dalia was away on a business trip. The gymnastics studio they sent their son to had an opening for him to practice one additional day a week. Crew looked at the number of hours their son was devoting to gymnastics and thought he was spending too much time at practice, while Dalia looked at that same number of hours and thought their son was not practicing enough. They couldn't see eye to eye and both discussed how exhausted they were from their three-day-long text argument. When situations are equivocal, we resolve them best by being precise, taking each other's perspectives, soliciting feedback, asking for clarification, and coming to agreement about how we'll move forward. All of that takes negotiation, which requires a lot of back-and-forth. That's hard to do via email or text.

The problem is that these two relatively lean media don't have great affordances for reducing equivocality. It's not super easy to interject to tell someone you don't understand, and the written word doesn't convey tone the way hearing someone's voice or looking at their posture or the heave of their chest might. Studies tend to show that teams and individuals who are assigned to make decisions about highly equivocal data in a lab setting do best when they communicate face-to-face, via videoconferencing, or via phone, and do worse when they try to reach consensus via email or some form of instant messaging or group chat. Interestingly, people take less time to make a decision via lean media versus rich, which suggests that they are not processing as many cues and that the absence of those extra cues leads to decisions that the teams are less happy with on the whole.

If good decisions based on equivocal data require precise communication, then lean media serve up another problem. When Jeanette and I were emailing back and forth about the personnel and budget issues, there was a lot of ambiguity about what our various actions meant. When she wrote "That's fine with me," I couldn't tell if she actually meant it or if she was just saying that to shut me up. When she didn't respond to one of my direct questions for two days, I couldn't tell if that meant she was mad at me or just busy. In her book *Digital Body Language*, Erica Dhawan provides numerous examples of how lean digital tools make it difficult to decode people's ambiguous words or actions—like not responding quickly to an email. As she concludes from her research on the ambiguity caused by communicating via digital tools, "People couldn't make heads or tails of the tone behind messages they were getting in emails, text messages, conference calls, and so on. Nor were they entirely aware of how their own messages were being received. More than just a glitch or a nuisance—*technology is such a pain!*—our shiny new communication tools were causing serious issues. Work and decision-making had slowed. Teams were in disarray. Employees were left unmotivated, distrustful, uncertain, and paranoid." Whew! Dealing with the ambiguity caused by our digital technologies is certainly exhausting.

But there's a solution: When we are faced with equivocal data or information that don't lend themselves to easy agreement, we must choose digital tools that will help minimize the inevitable ambiguities that crop up. When we know that the data or information we are working with is equivocal, we need to choose digital technologies (or analog meetups) that reduce ambiguity. I should have matched the equivocal data Jeannette and I were discussing with Zoom or the phone or another digital technology that would allow us to give each other instantaneous feedback, where I could sense more of her confusion, and where we could resolve misunderstandings immediately. That's also the decision that Crew and Dalia should have made to decide about their son's gymnastics practice. But matches work the other way too. It's equally as important to match nonequivocal data and information with lean media. That's because the research shows that if a decision is relatively straightforward, too many cues can surface related problems that stymie easy and straightforward decision-making. Think about the last time you agreed easily with someone but thought they gave you a side-glance or you heard a pause in their voice and it opened up an unrelated can of worms that derailed the easy decision. It turns out we end up much less exhausted if we also match tasks low in equivocality with leaner media.

CONVEYANCE OR CONVERGENCE?

A second matching consideration concerns the level of coordination necessary to make something happen. Not all coordination is the same. The kind of coordination we need to have with other people depends on how much our actions rely on theirs, also called "interdependence." Obviously, the more you and I depend on each other to program software or plan a night out with our mutual friends, the more coordination we're going to need. The less interdependent our tasks are, the less coordination is necessary between us to make it work.

Decades ago, the sociologist James Thompson developed a framework for thinking about different types of interdependence that still applies remarkably well today. Thompson suggested that we might think about our task interdependences with others as being either pooled, sequential, or reciprocal:

1. **Pooled interdependence.** Each person contributes independently to the collective outcome, with minimal direct reliance on the work of others. For example, a group of sales representatives in a retail company each operates in different territories. They work independently to meet their sales targets, which collectively contribute to the company's overall sales performance. Each salesperson manages their own client interactions and sales processes but relies on the same digital support systems, such as the customer relationship management software and product training provided by the company. Or in your personal life, you join a workout challenge with friends where each person tracks their gym time and logs it into a shared app. While each person's workout is independent, their combined efforts contribute to the group's goal, like collectively walking the distance equivalent to crossing a continent.

2. **Sequential interdependence.** The output of one person's work becomes the direct input for another person's task, forming a linear workflow. For example, in a content production team for an online magazine, a writer first creates an article, which is then passed to an editor for revisions. After the editor's approval, the article goes to a graphic designer who adds visual content, and finally, it is sent to a web technician who publishes it on the magazine's website. Each individual's work must be completed before the next person can begin their portion of the process. Or at home, one family member plans a weekly meal schedule and sends it via a meal tracking app. Another person orders groceries online based on the plan, and a third prepares the meals according to the sched-

ule. Each step is dependent on the previous one and requires coordination.

3. **Reciprocal interdependence.** These tasks require continuous back-and-forth interactions to complete the work effectively. For example, in a hospital setting, a doctor and a nurse managing patient care in an intensive care unit exhibit reciprocal interdependence. The doctor relies on the nurse to monitor the patient's condition and administer treatments, while the nurse needs the doctor's expertise to diagnose conditions and prescribe the appropriate medications. Their tasks are highly interdependent, requiring ongoing communication and adjustment based on patient responses to treatment. Or with your extended family, you're planning a reunion and you and your relatives use a group chat to discuss and decide on the date, venue, food, and activities. Each decision influences and depends on the others, requiring constant back-and-forth communication to align everyone's preferences and availability.

Generally, coordination needs are low when our interdependence with others is pooled or sequential. In situations of pooled interdependence, we don't have to worry too much about coordination because each individual or group works independently, contributing to the overall goal without direct dependency on the moment-to-moment activities of others. Tasks are relatively isolated, and interactions are limited to sharing common resources or adhering to overall guidelines. Sequential interdependence also requires fairly low levels of coordination because it's just the output from one individual or group that becomes the input for the next. This chain-like structure means that while each task is dependent on the previous one, the interaction needed to hand things off is still somewhat predictable and linear.

Digital technologies that Daft and Lengel would classify as lean are usually sufficient for managing the coordination of tasks that are pooled and sequential. These include emails, memos, or standard reports that provide necessary information without the need for immediate feedback. Since tasks are independent, the communication primarily needs to be clear and informative, not necessarily rich in transmission of cues. Detailed documentation and scheduled updates can ensure that each party knows exactly what is expected and when.

By contrast, tasks that demand reciprocal interdependence have the highest coordination needs. Face-to-face communication or richer digital technologies help manage the complexities and dynamism associated with these kinds of tasks. They require immediacy and a multiplicity of cues to handle the ambiguities and potential conflicts that might arise, ensuring that all parties involved can coordinate effectively, resolve misunderstandings quickly, and adapt to changes as they occur.

Think about whether whatever you're doing requires you to *convey* information to others, like you would if you were pooling or sequencing your tasks, or whether you need to *converge* on some shared meaning or mutual understanding like you would in a situation where tasks are reciprocal. Generally, if your coordination need is to convey information, use a digital technology that lets you send text-based information asynchronously or transfer documents or images that help to clarify things. When you're sending information to help someone understand what you've been working on or are trying to get them to pick up where you left off, that person needs time to process what you've sent. Technologies that don't demand an immediate response and remove nonessential cues that could distract from the key data are most useful in these cases. But if your coordination need is to converge on a common understanding or a collaborative output, use a technology that affords synchronous collaboration through which you can transmit multiple cues. When you need to discuss each person's under-standing of a situation, reach some shared agreement, and figure out how all the parts fit together, you need more active discussion and rapid feedback or else things stall out.

When we choose to use a technology that doesn't align well with the nature of our tasks, we end up with coordination problems that exhaust us. Using email or Slack for tasks that require constant, real-time collaboration can lead to frequent misunderstandings and the need for repeated clarifications, significantly increasing the cognitive load and stress levels of our team members. Deploying overly complex collaborative tools for simple, independent tasks can result in frustra-tion and a waste of resources. Both are examples of mismatches be-tween technology and coordination needs and will lead to exhaustion and decreased productivity or enjoyment.

YOU'RE SUCH A JERK!

After becoming good at recognizing what kinds of activities needed to be matched with which technologies based on equivocality and co-ordination demands, Mark thought he had it down. One of the para-legals who worked at his law firm stopped by his office to ask his opinion on a brief she was preparing. Mark said he'd look it over and get back to her. "I thought about the matching rule like you told me," he recounted to me on a phone call early one morning, "and I thought the best way to provide feedback was just to do it in Google Docs. The brief was pretty straightforward and she wasn't gonna need to come back to me to ask questions. So I just made notes throughout the doc for her." But in the days following his edits, Mark heard through the grapevine that the paralegal was unhappy. It turns out she was new on the job and thought that Mark's approach to work on the edits and comment on the brief in Google Docs was a signal that he didn't want to take the time to meet with her and mentor her. "I felt like crap," Mark reflected. "I was just trying to think about my exhaustion and hers of having a meeting that didn't need to really happen. I didn't think that there was some symbolism to the meeting that she was reading into it."

Mark's experience is, in my experience, common. The idea of matching is based on a rational notion of how we can optimize our information dissemination and processing to reduce digital exhaustion. But sometimes our choice of digital technologies is symbolic. A number of very persuasive studies in the media richness tradition have demonstrated that in certain circumstances, rational matching can lead to more harm than good if the choice of a particular technology sends the wrong signal about someone's intentions or motivations. If a certain type of communication or interaction in the workplace or among friends is symbolic of someone's love or care, or someone's indifference or annoyance, making a rational match around equivocality and coordination needs can backfire. Like me, you've probably been in a situation where you fired off a text or sent a quick Microsoft Teams message only to think "Hmm . . . maybe I should have called her up to tell her that" because you realize that it might have come off as insensitive to share the news asynchronously and via text, even if it was efficient.

Mark grew worried and stressed that he might have offended his new colleague. "I think I was more exhausted thinking about how I should have had the meeting than if I just had the stupid meeting," Mark said. When he realized the symbolism of doing that first meeting with the paralegal in person and that he had inadvertently signaled his indifference by making a rational match, he apologized to the paralegal and explained the situation. As Mark recounted, "She laughed and I could see the tension ease in her face. She told me that she was sorry she made a big deal about it, but that she worried that she'd been assigned to work with a jerk who didn't care about her. Kind of a bold thing to say if you ask me. But we laughed about it and it's all good." Mark's story illustrates a third consideration for the matching rule: Always make sure that your matching is not creating adverse signals. If it is, you may need to think about how to discuss the norms in your organization, friend group, or family around what acceptable forms of communication should be for particular types of tasks or issues.

Batch and Stream

There aren't a lot of great studies available that track statistics on real-time email use, but conservative estimates suggest that the average worker receives more than 120 emails a day and checks email nearly eighty times a day. All that email-related work adds up to nearly nine hours per week. And that's just email! While people are checking and sending emails, they also toggle back and forth between applications including Microsoft Excel and Word, and Google Docs and Sheets. Text message notifications keep arriving on their phones. They receive multiple Slack messages. They check their LinkedIn news feed at least twice during the day. This list could be a mile long. Data from Microsoft showed that users of its products spend more than half (57 percent) of their workday using digital communication tools and only 43 percent of their time using their so-called productivity tools. But let's just focus on email for a minute since there is the most research about that.

Jenn, who is vice president of accounting for a chain of fitness centers, has a higher total than the reported average. During the past year, she's averaged 168 emails per day. Jenn checks her email only three times per day—first thing after arriving at work, just before lunch, and about one hour before she leaves work each day. "I'm very disciplined about it," she remarks. "When I do my email in batches, I'm like a machine. I get them done quickly and then I have more time for focus on other tasks." Jenn reports a 1 on the digital exhaustion scale.

Nelson is also a vice president of accounting, but at a health food company. He has a slightly higher total than Jenn, averaging 184 emails a day. An analysis of his email logs shows he checks them about eighty-three times per day. "I start to get into something and I see I've got a new email, so I just check it. I even check emails when I'm in face-to-face meetings with people. I know it's rude, but I can't help it," he says. Nelson reports a 5 on the digital exhaustion scale.

Of course, email is just one of many digital technologies that Jenn and Nelson use and only one of several communication methods. Their behavior shows two approaches to handling information. Jenn is a batcher. She processes her emails in bulk at certain times of the day and does not check them outside of three prescribed windows. Her email behavior is an indicator of her behavior on other tools. She doesn't let herself get interrupted if she's working on a financial analysis. She also refuses to answer the phone or return a text if she's working on her email. She's very disciplined. She'll batch process those activities when it's their turn.

Nelson processes his email, and most of his other information, in streams. As soon as new information comes in, Nelson does something with it. If he gets a call, he takes it when it comes. If he receives a report, he reads it as soon as it comes in. And if he needs to manipulate data in a financial report, he does it just until the next job or piece of information comes in. He deals with that and then switches back. "I just can't ignore the lure of new information. And I like to deal with problems and issues right now," he says. Nelson's a streamer.

The terms "batch" and "stream" have technical roots. Originally, batch processing referred to a method of data processing by computers in which transactions are collected over a period of time and processed together in a single batch. This approach was standard practice in early computing when resources were limited. By waiting to batch process all transactions together, usually in a business's off-hours, limited computing resources could be given to employees during the day to deal with real-time demands. This method is still widely used today in many industries, such as end-of-day transactions in banking, payroll operations, and batch updates for inventory control.

The opposite of batch processing is streaming. Streaming involves processing data immediately as it becomes available, ensuring that the output is generated without delay. Think about the streaming services you use like Spotify or Netflix. Your phone or TV doesn't wait for the entire song or movie to download before it begins processing the data and rendering it into sound or picture. It begins dealing with the data packet by packet as it is received. That's why your song or movie sometimes glitches—it hasn't received the next piece of data yet, so it can't process it for you. Real-time systems are designed to handle tasks within a guaranteed time frame, promptly processing data so immediate actions can be taken based on the latest information. This is essential in applications where timing and immediate responses are critical.

Batching and streaming represent two complementary approaches to dealing with the deluge of data we experience every day. It's possible to batch or stream text messages, Slack DMs, phone calls, spreadsheet work, document editing, content creation, music listening, book reading, video game playing, and so much more.

We might be tempted to conclude that it's better to be like Jenn than Nelson—a batcher rather than a streamer. By just reflecting on what you've read so far, you would probably reason that batching al-

lows us to reduce context switching and conserve our attentional resources. Batching does reduce the demands on our attention, which is
a key driver of our digital exhaustion. We'll explore this more in a bit.
But Jenn's and Nelson's digital exhaustion scores tell only part of the
story. Even though he might have high levels of digital exhaustion,
Nelson's colleagues love him: "He's so available and he always gets
back to you right away," said one. And despite her lower exhaustion,
Jenn's colleagues are frustrated by her. "She will never make time for
you, and her answers often come after you need them," said a VP in
another division of her company.

Our choices about batching and streaming carry social consequences, particularly when we are batching and streaming in ways
that affect others. So unlike the first two rules, this rule is not definitive. I can't tell you that you should only batch or only stream. The rule
for this chapter is that you need to develop a strategy to batch and
stream in ways that reduce your digital exhaustion while maintaining
your reputation and reliability. The optimal orientation will almost
always combine batching and streaming behavior. Let's explore how
to make healthy decisions about what and when to batch and to stream,
what the division between the two behaviors should be, and why it
generally makes sense to expose yourself to a bit more exhaustion
than Jenn is comfortable with in order to maintain long-term success
and happiness in relationships at work and at home.

BATCHING

At its core, batching is a strategy to deal with interruptions. When
Jenn checks her email first thing in the morning, before lunch, and an
hour before leaving work, she is creating a system in which she artifi-
cially controls the unpredictable and untimely arrival of emails. Isa, a
chemist at a major drug company, batches her texts. As she describes

it, "When I'm at home, I only check my text messages at certain times. I'll check them before dinner, right after I put my kids to bed, and then right before I go to bed. I do it that way because I don't want texts interrupting my evenings with the family." Dan, a physician's assistant, batches his entries into the clinic's electronic health records system: "Some folks go to [the records system] right after every meeting and enter in relevant information. But that gets in the way of all the other things you've got to do. I block out two hours a day to enter my notes at once. That's all I do and people know not to bug me." Reggie, who works as an automotive engineer, batches the debugging required before he can submit his simulation models of car crashes to the supercomputer: "I'll put five or six runs in at a time. I wait till they all come back, and if they timed out, I'll just go in to them all one after the other at the end of the day and figure out what went wrong. I don't look at each one when it comes back, because that would just distract me from whatever I'm working on at the time." A key feature of our digital lives is that information comes to us on schedules that are not of our own choosing. We can choose to deal with it when it comes, or we can decide to wait and deal with it when we're ready. That's the idea behind batching.

In theory, batching is a great solution to exhaustion. By dealing with all of our emails, texts, or simulations at once, we can forestall the many context switches that steal our attention. But, unfortunately, the evidence on batching is mixed. In one study headed by Indy Wijngaards at Erasmus University Rotterdam, the research team conducted a field experiment with a large Dutch financial services firm. One group of employees agreed to batch their email, only checking and responding to messages three times a day—just like Jenn's practice. The other group maintained their usual email habits. After a month of email batching, the participants in the experimental group reported a major reduction in email interruptions throughout their day and, not surprisingly, a significant decrease in feelings of exhaustion. The effects of batching on reduced exhaustion were highest for people who handled a high volume of emails per day (more than twenty-five in this study). The control group, which made no change in their email use over the same one-month period, reported decreases in neither the number of email interruptions nor their level of exhaustion. It seemed that batching did indeed make a difference in levels of exhaustion.

But two weeks after the end of the intervention, the researchers found that only slightly more than half (53 percent) of the people from the experimental group continued to batch their emails. When they asked those who had stopped batching why they did so, they heard a uniform answer: Their coworkers and customers expected faster responses than batching would allow. Batching was a success, but peer pressure made it too difficult for many to continue.

In a study of smartphone notification batching conducted by Nick Fitz and colleagues at Duke University, participants who received notifications batched three times a day reported feeling more attentive, productive, and in control of their phones, compared to a control group who received notifications as usual. They also experienced lower stress levels and fewer interruptions due to phone notifications. However, participants who received no notifications at all experienced higher levels of anxiety and FOMO than participants in either of the other two groups. The no-notification participants were sensitive to the fact that the world was spinning while they were not checking their apps, and they grew fearful and anxious about missing out on important activities and letting other people down. Another study showed that batching may actually increase stress levels for those who are generally more prone to anxiety.

The general trend across many studies is that we get stressed out when we have to deal with data, information, or communications that come at us on someone else's schedule, not our own. Batching our work when using digital technologies lets us manage information processing demands on our own schedule and focus without interruption. This approach can reduce the immediate stress of reading, thinking, and responding, as well as our overall experience of exhaustion. But batching is not a panacea. Here are some ideas to consider if you think batching might be right for you:

1. **Batch if interruptions are constant and voluminous.** Batching works best if you experience many interruptions and they come at you throughout the day. Email again provides the easy example. If you're someone who only gets a few emails a day, batching probably isn't necessary. Or, if you are like Denise and your emails are concentrated at certain periods in the day, you might already have a natural batching system going. Denise works remotely on the West Coast of the US for a company headquartered on the East Coast. When she opens her email in the morning, she's already received between ten and fifteen emails from her colleagues who have been working for three hours. Denise batches those emails first thing, fielding the responses that come back right away as part of her batching sessions. Then she can do focused work for the next few hours. She returns to batching during her East Coast colleagues' final hours of the day, then she's able to once again do focused work for the rest of her afternoon.

2. **Group similar activities into batches.** When we think about batching, it's easy to slip into the mode of linking batch processes to particular digital tools. For example, batching our emails, edits on a report, Twitter reading, and so forth. That's one way to do it. But the research shows that a more effective way is to batch to-

gether groups of activities that are natural complements or that constitute a stream of work. Gloria Mark, the context-switching guru we met in chapter 1, advises batching groups of activities that are logically connected and require similar skills or that are aimed at achieving a specific goal. We tend to naturally organize our work and leisure activities into these groups as a way to manage the task complexity and reduce the cognitive load associated with switching between unrelated tasks. For example, let's say you're a teacher who is preparing for class. You might group together activities like reading an electronic article, updating a slide deck, emailing a guest speaker, and perusing the Blackboard learning management system to see what questions students submitted in advance. Those activities, though they involve using at least four different digital technologies, are a logical, coherent group because they all operate together to help us do one particular set of tasks. As we discussed earlier, our attention switching exhausts us more when we switch across domains and arenas than it does when we switch across modalities. Grouping together activities that demand similar cognitive processing and batching them even if they cross multiple tools is a helpful way to reduce exhaustion. The fact that research that focuses just on email batching doesn't find stronger reductions in exhaustion isn't surprising, especially if your emails are anything like John's: "When I'm on email or even texting it's like my head is in a hundred places at once. One email is from a customer, another is from a supplier, another is from my sister, another is from my plumber. I mean, I have to constantly stop and start and readjust and remember where I'm at and who I'm talking to every time I sit down to handle a bunch of emails." You do better if you think about grouping activities than batching just by technology type.

3. **Don't batch for too long.** Batching is about focus. Talk of focus and deep work is all the rage these days. And it should be—it's hard to get meaningful things done if we are constantly distracted, and we know that too much distraction is exhausting. Another reason not to batch for too long is that we often think that batching will allow us to "catch up." But research on how people batch email, texts, and more shows that no matter how long they spend on it, they rarely ever catch up. In fact, the more efficiently you get work out by batching, the faster data, information, and communication comes back, creating a Sisyphean cycle. When we think we have enough control to get caught up with some task and we fail, we become even more demotivated and exhausted. It's better to set a time limit on your batching to avoid fatigue from too much focus and exhaustion from the lack of control you feel over not clearing out your inbox, catching up with all those LinkedIn comments, or finishing all those edits.

STREAMING

People like Nelson who stream their digital communications by setting alerts and notifications to announce when new information arrives, and reading and responding to it as soon as it arrives, look very different from batchers like Jenn. Their patterns of data processing and response are instantaneous. If you're not a streamer, you've certainly noticed one. They're the folks who seem to comment almost immediately on your Facebook post, they're the first person to put a heart emoji on a text in a text chain and they respond to your email mere seconds after you hit send. Streamers are on top of their data sources and desire quick action.

Their digital housekeeping also looks very different from that of batchers. Most batchers strive to respond to as many messages as they can during their batching sessions. Their goal on email, for example, is to get the inbox to zero. Of course, most never achieve that goal, but they try to, in spirit if not in practice. They file or delete messages as soon as they send them, and they generally equate a clear inbox with a clear mind. Streamers tend to have large and unwieldly inboxes. They don't delete most messages. When Nelson showed me his inbox, he had 8,746 emails in it. 1,452 were unread. When I asked him why he had so many unread emails, he said, "Every time something comes in, I scan it quickly. If it's not important I just ignore it. No biggie." "Do you ever plan on cleaning your inbox out?" I asked him. He looked at me with a puzzled face. "Why?" he asked me in earnest. When Jenn showed me her inbox, she had forty-three messages in it. All of them were marked as read. "I keep thinking I'm going to get back to these final messages, which is why they're still sitting here," she told me. "But other important stuff keeps popping up. I just don't have the time. None of it is mission critical, but I should respond. It bothers me to see them sitting there."

Streamers like Nelson and batchers like Jenn orient to their communication environments very differently, and they each have their own preferences for how to process information. A study by Laura Dabbish and Robert Kraut, both from Carnegie Mellon University, found that for streamers, restricting email checking to specific times during the day (what batchers do) was associated with feelings of exhaustion, but checking and responding to emails right away was not. Individuals who received larger quantities of email in general reported more exhaustion than those who received few emails. Yet the researchers concluded that for people who prefer streaming, checking messages right when they arrive may actually help to reduce overload. That's because only checking messages at specific times can cause them to accumulate, leading to more emails to handle at once, which can make people feel anxious and angry. For people like Nelson, the thought of having to deal with emails piling up was dreadful: "I hate when there are too many emails to check. That's why I just do them when they come in. It's much more pleasant that way."

Streaming can also provide a sense of control over one's work. Batchers find control by dictating when they will work on something, and they don't let the influx of new messages or data determine their priorities or patterns of attention. Streamers find control in accomplishing things. For many streamers that I've interviewed, it doesn't matter whether those accomplishments are big or small. What matters is that they happen. As Jodi, a streamer, once told me, "So much of my job is me doing things that just take a long time. You make little wins now and then, but that's it. I'm an anxious person, so I just feel better when I get things done. So if I can give you quick feedback on your presentation or run a quick analysis of new customer usage pattern data that just came in or respond to your email, I feel like I'm getting stuff done. I need that." Breaking up her concentration on larger projects to stream incoming data, rather than setting aside blocks of time to do so, made Jodi feel like she was in control.

There might be something to the fact that Jodi refers to herself as an anxious person. One study showed that anxious people were more likely to stream their email on days they experienced high workplace anxiety, and that they performed better when they streamed. The authors suggest that the findings on streaming "point to an adaptive function of work email activity—it serves as a regulatory behavior that enhances performance outcomes on days employees feel anxious." In other words, if you're an anxious person and your day is making you feel especially anxious, streaming your email may help you to feel in control of your ability to get things done and enable you to do more.

Taken together, these findings suggest that streaming has several positive benefits for people who live with anxiety:

1. **Immediate task engagement.** Channeling anxious energy into doing immediate and burning tasks—like responding to emails or writing a quick summary for someone right after the request comes in—can help you regain a sense of control and direction. Streaming provides clear, action-oriented responses that can help redirect focus from anxious thoughts to productive activities.

2. **Clear communication.** Engaging in more frequent email communication can help ensure that you are on top of your responsibilities, reducing worries about missing out or falling behind on important tasks. This proactive approach can alleviate anxiety by reinforcing a feeling of competence and accomplishment.

3. **Active problem solving.** By increasing streaming activity, anxious employees can be more actively involved in problem-solving and decision-making processes. This active engagement and solution-oriented thinking can help them feel reassured rather than ruminate on what could go wrong.

4. **Feedback and support.** Streaming can also increase the frequency of feedback and support from colleagues and supervisors, which can make anxious employees feel less isolated and stressed, especially when tackling difficult tasks or situations.

The general evidence seems to suggest that for most people, batching is more effective than streaming for keeping exhaustion at bay because it reduces the need to context switch and creates opportunities to do deep work. But streaming may be more effective for people who experience moderate to high levels of anxiety because it can give them a sense of control. It's clear that deciding to either batch or stream exclusively would be ill-advised. Instead, figure out what mix of the two activities will work best for you, and then build a plan about how and when you'll do each. The key is to be intentional. Make sure that your choice reflects your own personality as well as the demands of your work and personal connections, and that it gives you enough of a sense of control and focus that you won't exhaust yourself.

Wait. One Hour.
One Day. One Week.

Jed is an analyst for an investment bank who prides himself on his fast response time. He told me, "When people send me messages, I've got to respond right away. They need something, and I've got a good reputation at getting back to people fast and helping them get things done." Jed's speedy response time on Slack, email, and via text at work was only the tip of the iceberg. "Everyone deserves the respect to get their questions answered fast, or whatever they need," he continued. "If someone posts an interesting post on LinkedIn, I always like it or comment as soon as I see it. I'm dependable, and I take pride in that. And it's the same on Instagram. Boom. Don't think about it, I just do it." Jed's description of his response behavior on digital tools is filled with interesting words—Reputation. Respect. Dependable. Pride. Boom.—and even one catchy corporate slogan, "Just do it." I say "interesting" because Jed's approach to his responses seems to be as much geared toward shaping how others view him as it is to actually providing relevant and timely information and feedback to others.

Because I was doing research at Jed's bank for a few weeks, I was able to talk to some of the colleagues he prided himself on responding to so quickly and a couple of friends with whom he had lunch regularly. I asked them what it's like to communicate with Jed. Here are three responses: One from Peter, another analyst at the bank. One from Rebecca, Jed's boss. And one from Marissa, Jed's friend who works for a clothing distributor several blocks away.

Peter: Jed's a great colleague, always super helpful.

Me: How would you characterize his communication style?

Peter: I don't know, normal. I mean he gets back to you if you need him to.

Me: Is he especially quick or helpful?

Peter: No. I mean, sure, he's helpful, but just sort of normal, I guess.

·····

Rebecca: Jed is a wonderful employee. He's friendly and gets the job done. He's got the knack for being a good analyst.

Me: How would you characterize his communication style?

Rebecca: Do you mean does he communicate appropriately for our business? Sure. I would say it's good.

Me: Is he especially quick or helpful?

Rebecca: In what way? If I need an answer, he'll get me one. Generally, he's fine. If I had to give him some advice, though, I'd say that sometimes he seems to respond too quickly without thinking things through. Not everything is super urgent and he could take a bit more time to be more comprehensive.

·····

Marissa: Jed's a good friend. We've known each other since college and it's fun to work close by so we can hang out.

Me: How would you characterize his communication style?

Marissa: I guess maybe I'd say he's always on his phone? Like you message him and he always messages you right back?

Me: Is he especially quick or helpful?

Marissa: Yeah. Like I was saying, he's real fast. I mean, sometimes I'm like, "You didn't need to get back to me so fast or like my socials every single time." I don't really expect a response that fast. He must have more important things to do than get back to me fast all the time. Right?

After conducting these interviews and five more with additional colleagues, I felt bad for Jed. He clearly went out of his way to respond to people quickly, believing that others valued his promptness and building his self-worth around being someone who was there for you at any time. Although each of the colleagues and friends I spoke with had good things to say about Jed, none of them lauded his short response time. Most didn't even notice it, and the ones who did either wondered why he was prioritizing them over the many other things he had to do or thought that he might benefit from slowing down. Perhaps the most unfortunate thing is that working to be a fast responder has led Jed—and the many others I've talked to who operate in similar ways—to become exhausted. You probably won't be surprised to learn that Jed scored a 5 on the digital exhaustion scale. And he understood why. As he told me,

> It's really tiring to make it a priority to get back to people quickly. I have to start and stop what I'm doing [*attention switching*] and I get anxious [*emotion*] if I don't know the answer or I think other people are wondering why I haven't gotten back to them [*inference*]. I can feel it at the end of the day [*Level 1 exhaustion*] and I've noticed that I'm just not quite as good as I used to be at staying on top of everything and I can feel I'm definitely getting more worn out as I progress further in my career [*Level 2 exhaustion*].

Could there be a more textbook case of digital exhaustion? Still, Jed doesn't think it's cool to wait.

The reality is that even though we might want to respond quickly, most people don't need such fast responses. Rebecca, Jed's boss, didn't need him to respond so quickly, and even suggested that slowing down might make him *more* competent. Marissa, Jed's friend, didn't expect such a fast response from him either and worried that he didn't have his priorities straight. And Peter, Jed's work colleague, couldn't even recall whether Jed responded quickly or not. Laura Giurge of London Business School and Vanessa Bohns at Cornell identify what they call an "email urgency bias," an "error" they say "is the result of an egocentric bias that leads receivers to overestimate senders' response speed expectations." The authors tested the email urgency bias across eight different experimental studies involving slightly more than four thousand people. Their results showed that in general, email recipients viewed messages to be more urgent and require faster responses than the email senders did. Perhaps most importantly, the more urgent recipients felt an email was, the more stressed they felt about having to respond to it and the lower they rated their own well-being. When those emails came outside of work hours, the email urgency bias was even stronger and recipients grew more stressed at the thought of having to respond to them. Other studies have examined the relationship between perceived email urgency and exhaustion and found that the more people believe emails they receive are important and need to be responded to quickly, the more they report feeling email overload. It turns out that we often think the communications we receive are more urgent and need a faster response than the senders do.

I've met a lot of people like Jed. We live in a world in which speed is lauded, busyness is praised, and quick action is heralded as virtue. Jed and those like him are products of our culture. Not only do they want to fulfill their perceived mandate to be quick to respond, they also see doing so as an opportunity to distinguish themselves from others on an attribute that we seem to widely admire. Just as people began recovering from the pandemic and readjusting to a new balance between in-office and remote work, famed organizational psychologist Adam Grant published an op-ed in *The New York Times* titled "Your Email Does Not Constitute My Emergency." In it, he argued that our expectations for fast response times via email and other communication tools are vastly exaggerated and inappropriate in today's digital world. As he wrote, "For most of human history, being responsive meant paying attention to the needs of a small group of people in your immediate vicinity: family, friends, neighbors, colleagues. Now there's no limit to the number of people who can barge into your inbox, ping you by text, and slide into your DMs. Digital overload cries out for us to redefine what it means to be responsive. The true test of a relationship isn't the speed of the reply. It's the quality of attention you receive." And with characteristic wit and wisdom he advised, "How quickly people answer you is rarely a sign of how much they care about you. It's usually a reflection of how much they have on their plate. Delayed replies to emails, texts and calls are often symptoms of being overextended and overwhelmed. If the message isn't time-sensitive, we should count delays in weeks or months, not days or hours."

The week after publishing the op-ed, Grant posted about it on LinkedIn. The comments were illustrative. Anyone who follows Grant's social media posts is used to seeing an endless scroll of affirmations and praise of his recommendations. But this time was different. The comments were largely critical of his message, and argumentative to boot. One commenter wrote, "I disagree . . . it IS rude to wait a week to reply to a message." Another commented, "I understand waiting 24 to 48 hours, but if you're taking a week to respond, you're ghosting, which is highly unprofessional." Another was far more blunt: "This is stupid advice and out of touch with the real world." A slightly more reflective commentator wrote, "Everyone is trying to be the first to respond and claim credit, and sadly it works." Such commentary shows just how strong and deeply ingrained our cultural expectations around quick responses are. That means that, more than ever, we need to be mindful and reflective about how and when we respond to the messages that our digital tools make so readily available. That's why this rule is to wait. If we want to reduce our exhaustion, we would do well to slow down.

TRIAGE THE DIGITAL EMERGENCY ROOM

Our digital tools have created frictionless communication. It's all too easy these days for anyone to send us a message about anything they desire. They don't need to make the calculation about whether their communication need is great enough to invest the time and effort to hop on a horse and ride to us or go to the store and buy stamps to post a letter. Instead, they think it, text it, and now it's in our queue demanding our attention and action. That means that in today's digital workplace, and in our digitally dependent home lives, we have more communications than ever, but no more resources to deal with them. If you've ever had the urge to send someone a link to LetMeGoogleThat .com, you know what I'm talking about. As these requests pile up, they sap our attention, raise the specter of attributing intentions, and spark feelings of guilt and anger. With more inbound communications and limited resources, we can't respond to everything immediately or we'll burn ourselves out.

That means we have to triage our inboxes, text chains, and social threads. Triaging involves assessing all available cases of something, comparing them to one another, determining which is most urgent or important, and creating a prioritized list of actions. You've most likely heard this term used in hospital emergency rooms, where nurses must decide in what order patients should be seen by physicians. They triage by prioritizing gunshot wound victims over children with ear infections. Our inboxes, social threads, and messaging platforms can feel like an emergency room. Each incoming message is making a case that it is the most important thing you should deal with. But in reality, you have to evaluate which messages you'll prioritize. That's difficult to do. We need more and better information about the urgency of the communications we receive to triage our responses effectively. This is where waiting becomes useful.

My first exposure to the idea of waiting to respond came in graduate school. Like Jed, I was eager to please and prided myself on being a "good communicator," which for me meant fast responder. But while I was getting my PhD, I worked as a teaching assistant in an undergraduate class on organizational behavior taught by a very seasoned professor and learned an important lesson about responding too quickly. About two weeks into the class, I was feeling overwhelmed by all the emails and messages from students with questions about the course assignments and the concepts we were covering. It felt like I was spending half my day responding to emails from 120 students. The emails came at all times, interrupting my reading and research. I grew angry and made inferences that these students didn't care about me or my studies. I was exhausted. I mentioned my frustration to the professor and asked if she had any advice. "Oh yeah," she said. "Make sure you wait one day before you respond to their emails. Seventy-five percent of the problems will take care of themselves."

I wasn't sure what she meant. How would the problems take care of themselves? And wouldn't the students be annoyed at me? They were paying lots of money for this class, and I wasn't an esteemed professor for whom they would patiently wait out of respect. I was the lowly TA, just a couple years older than most of them. But I tried it. I waited one day before I responded. She was right. Several hours after receiving a demanding email, I would receive a second email from the student: "Never mind, I got the answer from one of the other students." "Sorry I sent that note. I should have checked the syllabus. I found the answer there." "I thought about it more and it's not really that big of a deal. No need to respond." I wouldn't say it was 75 percent, but close to two thirds of the inquiries I received were rescinded.

This was my first lesson about the power of waiting: Often, many of the urgent requests we receive will triage themselves. Years later, when my research team and I spent three months observing emergency pediatric transfers between hospitals in the Chicagoland area, I saw that real patients triaged themselves too. We were working on a project with Children's Memorial Hospital (since renamed Lurie Children's Hospital) to use digital technologies to improve the transfer of data across hospitals in the region. The hospital was struggling to accommodate the number of kids who might need transfers. One hospital administrator explained to me that Chicagoland's population had grown 20 percent over the last decade, "But we don't have any more children's hospitals now than we did then." As one nurse who was particularly skilled at facilitating transfers explained, "Our job is triage. Sure, we would love to help every sick kid here at Children's, but we have limited capacity. We have to get the right information to be able to figure out which kids are in most need of advanced help. Then we prioritize those and initiate a transfer."

Magda, one of the nurses, told me, "One thing you learn is that if you hear from a doctor [at a community hospital] that a kid is really sick, but you look at the data and he doesn't look that sick to you, your best bet is to just wait for a bit to see what happens. In our experience, a lot of the time the kid gets better and the doc tells you it was a false alarm." Thinking back on my TA days, I jokingly asked her if "a lot of the time" meant 75 percent. "Probably about that, actually," she said to my surprise.

What the transport nurses knew about the pediatric patient transfer process is what Giurge and Bohns learned about email communication: It was easy to get a message from a concerned doctor and believe that the situation at the community hospital was more urgent than it actually was. What made the transport nurses so effective was that they developed a process to help reduce the urge to exaggerate the perceived urgency. They examined the data dispassionately, thought critically about the symptoms and their severity, and reflected on the communicating and data-sharing tendencies at referring hospitals (and the doctors at those hospitals). Then, they created their own list of urgent priorities based on the hospital requests and responded to them. Along the way, many problems resolved themselves, like when a hospital would call and say the child's symptoms were abating quickly and they no longer needed a transfer. The transport nurses were responsive to urgent cases and responded quickly, but worked more slowly on others.

We can adopt this triage strategy to deal with the immense influx of communications made possible by our digital tools. The trick is to balance a measured view of the urgency that other people attribute to their messages with the urgency that we attribute to them, relative to the other tasks in our portfolio. It may sound cold and calculated, but in hospitals where lives are at stake, triaging is an effective algorithm to match limited resources with the most important problems. You may think that triaging feels rude or disrespectful. But as we've discussed above, most messages we receive are not as urgent as we think, so a slower response usually won't have negative consequences. And as Erica Dhawan notes, being responded to more slowly can actually feel sort of humanizing and encourage people not to send so many messages to us in the future. As she writes, "Being triaged might not feel much better than being ghosted if you have an urgent question for your boss, client or colleague . . . I've found that it forces me to confront my own main-character syndrome—the idea that we all play a starring role in the movie that is our life, with everyone else merely the supporting cast. It makes me acknowledge that the 'ghosts' are, like me, full, complicated people with off-screen demands that might often pull them away from digital conversations."

BE THE CHANGE YOU WANT TO SEE

Supriya is a mechanical engineer for a large aerospace company. She is a very strong proponent of waiting to respond. As she told me, "I've noticed over the years that when I respond to someone quickly, they respond back quickly. It happens on email for sure, but also on social media or just putting comments in a Google Doc. If I respond more slowly, people also respond to me more slowly and usually not as often." Supriya's observation is backed by good science. A number of studies have shown that we tend to entrain to other people's patterns of response. That is, we move in a tightly choreographed dance. The faster they respond, the faster we respond. And the faster we respond, the faster they respond. We keep waltzing round and round.

The pattern holds across multiple modalities, including email, text, and instant messaging. One research team found that in their corpus of sixteen billion (yes, with a "b") emails exchanged by two million users, the variable with the "highest predictive power" in explaining someone's response time was "the median reply time of the replier from earlier replies." In this study, the "replier" refers to the person who sent us a message. That means that our response times are dictated in large part by the response times of others—others who are responding to us based on our past response times. Supriya's smart move was to stop the music, restart the dance, and take the lead. As she told me, "When I finally just stopped responding to people based on how quickly I thought they wanted me to and started doing it on my own time, my life got so much better. Most people actually started mirroring my behavior and they responded slower too, which meant they also responded less often, so I get fewer messages overall. That has totally helped me feel less anxious and less exhausted."

The other benefit of responding slower is that we can more fully process and make sense of the information others send to us. When we respond quickly without fully digesting what other people are saying, we act like the students I TA'd: We ask questions we don't need to because the answers are already there. Many studies show that if we delay our response, we process better and more fully and we ask for fewer clarifications. This pattern of behavior is so strong that even studies of preschoolers show that delaying and thinking before responding improves our performance on all manner of tasks. If we take our time, we reduce the communication burden not only for ourselves but for others too. Like Supriya, we can take the first step to create a new cycle of entrainment by purposefully slowing down the speed of our outgoing communications, which can then slow the speed of incoming messages, thereby reducing their volume and tendency to exhaust us.

GO BIGGER AND LESS OFTEN

That study of sixteen billion email messages revealed another important finding: In general, the more emails people receive, the faster they respond to them and the shorter their replies. As we've learned, that's not a great pattern for reducing our exhaustion. But it gets worse. The data also showed that shorter emails are typically responded to more quickly. That means that if we respond quickly, we're more likely to respond with shorter messages and our shorter messages will trigger people to respond to us faster than if we sent longer messages. Yet another way for us to get trapped in an exhausting cycle.

Here again is where waiting can be useful. The longer we wait to respond, the more likely we are to send longer messages. The longer our messages are, the less likely people are to reply to them with anything more substantive than an acknowledgment of thanks. And if they do respond, it's likely to be with a similarly longer message with a slower response time. The data show this clear pattern. But why?

Marcel, a community relations manager at a large government research center, has a theory he claims has served him well throughout his career. He told me that he likes to send fewer messages to people that are more comprehensive: "I don't do quick responses. I typically wait to respond until I can really reflect and craft something meaningful and something that's useful to other people. When I'm more deliberate, people just respond less and there is less back-and-forth because all the detail is there. It's kind of like one and done. That really reduces the amount of communication I have to do overall and reduces the overload I feel." Marcel's theory has two parts. The first is that most digital communications make it possible to reflect and carefully craft his remarks in ways that real-time, synchronous tools don't. He can take advantage of these affordances to be more "meaningful" and "deliberate." The second is that the comprehensiveness of his response reduces the need for other people to respond to him in short bursts for more clarification. Let's see how Marcel's theory holds up to the evidence.

First, let's look at the affordances of asynchrony. My University of California, Santa Barbara (UCSB) colleague Joe Walther has spent his career examining how people can use digital tools effectively to maintain relationships with others. He argues, in line with Marcel's intuition, that digital tools that afford asynchronous interaction can actually help us be better communicators than we can be face-to-face or by using synchronous tools because we have more control over our responses. As he writes, "The capacity to change the content and appearance of a message before it is emitted, or abort a message and begin anew, is a luxury not afforded by FtF [face-to-face] interaction." His studies have found that when people are able to delay their response times so they can think through and carefully edit a message, they report higher satisfaction with their response, their responses tend to be much more comprehensive, they are more mindful of their recipient's needs, and they tend to generate more relational intimacy with their recipients. In short, people who delay responding feel better about their own performance and more effectively take into account the needs of others. These findings hold across all manner of asynchronous digital tools, including email, text, instant messaging, and social media. The evidence produced by Joe and many others in support of his "social information processing model" gives credence to the first part of Marcel's theory that waiting to respond produces better, more useful responses. The second part of Marcel's theory is that these better responses will forestall the need for lots of back-and-forth. Here the data also show strong support.

Another big-data study—this one analyzing a measly eight hundred million emails exchanged among one hundred thousand people—found that shorter emails were deleted by a receiver quickly (often within five minutes of receipt) while longer emails were held on to for much longer (in many instances more than a week) and were often revisited—25 percent were revisited only twice, and 64 percent were revisited up to five times. The researchers supplemented their email data analysis with surveys of more than four hundred people to determine what folks did when they revisited emails. Seventy-four percent of revisits to emails were to find information, and only 20 percent of revisits were to take an action such as responding. Of those revisits to find information, 25 percent were to get answers to a question they asked or to solve problems, 24 percent were to get instructions to perform some task, and 22 percent were to review an attachment with detailed explanations. What is important about these findings is that longer emails served as founts of knowledge and information over a long time horizon. People could go back to them and mine them for useful information without having to reach back out to the email sender for updates or more clarifications. Thus, supporting the second part of Marcel's theory, taking the time to send one longer email with more detailed information carefully crafted for the recipient can pay major dividends on the up-front investment of time. As one of Marcel's colleagues told me, "When Marcel sends me something it's really good. I can use it as a reference and I rarely have to bug him again. I have lots more back-and-forth with most people to get clarifications on things than I do with him. He's definitely an energizing colleague."

CALCULATE THE TRADE-OFFS

The choice about how long to wait to respond to the deluge of communications and data we receive depends on many factors, and only you can really decide what is appropriate. I find it helpful to think in terms of response intervals of one hour, one day, and one week. As I've applied the rule of waiting to my own life, I've found that looking at every inbound message and triaging them into these three categories is helpful for thinking about how much effort I will devote to each response. I've become a staunch triager, and I combine my triaging with batching.

I empathize with Jenn, who we met in our discussion of Rule #3: I wish I could close out my day with a completely empty inbox. That sounds glorious. I'm never even close. For many years I only checked my emails, texts, and social media accounts at specific intervals throughout the day, and I determined what I would check at those times based on my triaging schedule. I experienced a lot of benefits. Those several years that I was a meticulous triager and a dedicated batcher were probably the most productive of my professional career. I published a large number of academic papers, consulted with a wide array of companies, and did some of my best research. I could feel the energy I got from intense focus and deep work, and I experienced very low levels of exhaustion. Among people who make a living thinking about technology and its effects on our work and life, I'm not alone as a slow responder. Cal Newport professes a similar practice: "I'm bad at answering e-mails. I sometimes go a whole day without looking at my inbox (and sometimes even longer). I ignore messages. People I know well tend to call me when they really need to know something. I'm not bad at e-mails on purpose. If anything, I'm apologetic and ashamed about it and try to be more responsive when I can." Waiting might not be popular, but it sure can be productive and stop our exhaustion.

While waiting has many positive benefits, here are some email and text messages I received during this period in which I took waiting to its extreme and batched exclusively:

- "Hey Paul, so did you get this message? It's been awhile and I haven't heard from you."

- "We're going to cancel this offer because we haven't heard back from you."

- "Dude, why don't you like me anymore?"

- "I wish you would have gotten back to me this morning, I had to find another way to get it."

It's kind of painful to read those messages now. In my zest for productivity, I made others bend to my timelines. I learned from trusted colleagues that I had a reputation of not being reliable because I wouldn't get people answers immediately, and my friends grew annoyed because I was difficult to coordinate with. Some of my family members told me they wondered if I didn't like them. One of my PhD students once said to me, "You've got a definite pattern. You respond to your emails in batches. I know that if I want to get your attention immediately, I just email you when you're batching, then you'll respond." I asked him if that was annoying, and he said, "A little bit. But at least I know the pattern and I can work with it. I know that's how you get so much done, though." My extreme triaging and batching were reducing my digital exhaustion, but it was coming at a reputational cost.

Today, I still practice triaging and I'm still a batcher. But I'm much more fluid about it. I do the bulk of my information processing and communication in chunks. I also make sure to send out notes to tell people to expect a delayed response if I've triaged them into anything longer than my "one day" category. That strategy is also based on re-search showing that setting expectations about response time makes both the expectation setter and the expectation receiver feel good. I group together similar tasks that don't require too much context switching, and I still have set times when I respond to most of my emails and texts. I respond with slower messages that are more detailed, and like Marcel, I find that I get fewer requests for clarification and fewer additional questions that will steal my attention at some later time. But I also stream more than I used to. I monitor my communication channels at periodic intervals and triage messages and tasks, respond-ing to ones that need immediate action or handling quick edits or data analysis right away if it appears urgent. In some ways, I'm less ex-hausted because I'm also not worrying so much about letting others down. Triaging, combined with limited and intentional streaming mixed in with my batching, goes a long way to reduce my exhaustion while allowing me to be responsive to others.

Don't Assume

lonso works at a landscaping company in California as the foreperson of a team of forty. His company doesn't provide him with any technology for coordinating his workers. But his team is often working at a dozen locations at a time, so he relies on texts, phone calls, WhatsApp, Facebook Messenger, and even Instagram to keep tabs on everybody and solve problems. He constantly sees his workers posting to Instagram while they're supposed to be pouring concrete, and they often don't respond to his texts in a timely manner. When he arrives home from a hard day's work, he's physically exhausted from the manual labor and mentally exhausted from using so many tools to coordinate his work. While he's trying to relax, he sometimes sees posts of coworkers out for drinks without him. All of these data inputs lead him to make assumptions that exhaust him further. As he laments, "I know I shouldn't jump to conclusions about the things I see my guys doing and how it affects their work. But most of the time I just can't help it. It's all right there in front of my face."

Our digital world fuels our propensity to make assumptions.

When we see people recount exploits from their weekend on Facebook, post beautiful pictures of themselves on Instagram, or describe the amazing sale they just closed in Microsoft Teams, we see only a piece of their lives. Although we know there is much more behind the scenes, a strong body of research shows that the digital environment compels us to hyperfocus on what is in front of us and use that information to form assumptions of others—assumptions that are often ill-founded.

The study of assumptions occupies a central role in the field of social psychology. For example, it is the key mechanism at play in the well-known fundamental attribution error. When we see evidence of people doing bad things, we tend to assume that their bad behavior is caused by some innate disposition. We attribute their behavior to them just "being" bad. But when we do something bad, we assume we wouldn't have done it were it not for some external influence, attributing our own behavior to situational circumstances. Assumptions also play a starring role in the ladder of inference, a process that explains how we arrive at conclusions about others:

- We are exposed to limited data about people's behaviors or actions

- We focus on that data

- We interpret the data in light of our own worldview

- We make assumptions about people's dispositions, actions, emotions, motivations, and more

- We draw conclusions about them

- We form beliefs based on our conclusions

- We take actions based on our beliefs

Both models of human behavior treat assumptions as the result of more or less automatic processing. We typically make assumptions automatically, without realizing we are doing so, and we begin to feel a certain way about someone without exactly knowing why. Stopping ourselves from assuming things about others based on the glut of data made available about them through our digital tools is hard. Here are some ideas that my research shows can help.

DON'T OVERINTERPRET WHAT YOU CAN SEE

When Tim joined a large government-funded research center in Colorado as the new director of information technology, he found a mess. The IT technicians under his management were scattered across the various labs in the research center. They didn't talk to one another to share best practices or learn. They were all generalists. If you had a printer problem, your local IT person would fix it. If you had a cloud storage issue or problems with the software integrations for one of your highly calibrated instruments, the same IT person would help you with those too. Tim saw an organizational design problem: lots of smart people who were acting as generalists but not building the specific knowledge required to deal with the particular problems faced by scientists and administrators across the research center.

To fix this problem, Tim reorganized all the IT techs into one group and gave them a new digital project management tool that enabled visibility into what everyone else on the team was working on. The idea was that if an IT tech could see that someone else had successfully solved a specialized problem that was very tricky, that tech could ask them for help and assistance. Tim also hoped that as the IT techs developed specific expertise, they'd get assigned jobs that reflected their expertise, and over time they'd be solving complex problems easily.

The first couple of months showed no real change. But slowly, the IT techs did more projects and entered more documentation about how they solved them into the project management tool. Nobody was working any differently than they had before. But the digital tool made important data points visible that were previously invisible to the team. Before they started using the tool, few IT techs had any clue what jobs their colleagues were working on. The tool made job assignments and completion time easy for everyone to see. The IT techs also didn't really know what other people did to solve a user's problem. The documentation that they now entered into the tool made it easy to see what steps they took to solve the problems they'd been assigned. Aspects of the IT techs' work had now become visible to their colleagues, and guess what? Those colleagues started making assumptions. They began to assume that because people had done certain jobs in the past, they would want to do those jobs again in the future. They also assumed that if someone wrote some lengthy documentation about a solution, they solved the problem well. The visibility of certain kinds of data fueled their assumptions, and they acted on them by starting to assign their colleagues jobs similar to those they had previously completed quickly and had written documentation about.

Most of the technicians were furious. As Bridgette, one of the IT techs, told me after seeing she'd been assigned yet another printer fix: "I've never been so annoyed at my job. It's like all my coworkers just think I'm 'the printer bitch.' They keep assigning me all these stupid printer jobs. I don't want them." But Bridgette had a secret and somewhat devious plan to fix the problem: She decided to game the system. As she told me, "I'm going to fix this printer issue like I always do. But I'm going to start writing really shitty documentation. I'm going to make it look like I didn't really do a good job or know what I'm doing. If I do that a few times, people will start to think I'm not so good at printers and I'll get assigned something else." I saw far more egregious moves than what Bridgette planned as I observed these technicians for several months. Some would get a job in an area they really wanted and not know how to solve the user's problem well. When this happened, they would google some documentation from the web that they thought looked really good and paste it into the system as though it were their own. "That'll get people to think I totally nailed this one," the perpetrator would say to me unabashedly as I watched him do it.

The problem the team faced was that the new digital project management tool made a few data points about their behavior visible, like what jobs they'd done, how long it had taken them to do it, and their documentation of what they did. But it did not make many other data points about their behavior visible, like whether they solved the problem to the user's satisfaction, whether they liked doing that kind of job, or whether they provided good customer service. In their job assignment routine, the IT techs overinterpreted the visible data, which, as we saw, led to unhappy workers and, ultimately, a database filled with inaccurate information.

I helped Tim and the team make two big changes to solve this problem. First, we discussed the consequences of using limited visible data to make assumptions about the motivations, skills, and desires of others. The funny thing about assumptions is that they are difficult for their owners to recognize. So even though Bridgette understood that her colleagues were using the data they could see in the project management tool to assign her jobs, she didn't recognize that she was doing the same thing to others. As Benito, one of Bridgette's colleagues later told me, "It was eye-opening to think about the assumptions I was making about everyone that I didn't realize. And now I see too they were doing the same thing as me. It was always exhausting thinking that everyone was trying to give me all these crappy jobs they didn't want. But now I get that they weren't doing that so much, but they were doing like I was and just not thinking about why they were making decisions." The second change was to make some modifications to the project management tool itself to capture new kinds of data that actually mattered to the team. We built a module that allowed the IT techs to rate their interest in jobs and we used a function in the tool where they could enter information about things they were learning and challenges they had that weren't directly related to any particular job. Adding both these kinds of data helped create a richer set of options and interests for people to use as a basis for assigning jobs in the future—and to act as a useful backstop for when they inevitably lapsed and mindlessly drew on the data visible in the tool to make their assumptions about who was best suited to do what.

A very different context in which we are prone to make assumptions based on limited data is social media. As discussed in chapter 2, people like Dean who see Instagram posts of their friends' bike trips through Europe don't see the metaphorical full picture. The curated nature of social media leads us to overinterpret the visible in a big way. Perhaps this is why research has shown that social media is a playground for cognitive biases to run amok. For example, other people's selective sharing on social media activates the availability heuristic, where the things we're exposed to and have immediately at hand shape our understanding of someone's life, even though we are neglecting the broader, unseen context. Social media also kicks off a recency bias, wherein recent posts are overemphasized (in our minds and also by the platform's algorithms), making it seem like these moments we are seeing now define someone's entire life. And, of course, a spotlight effect comes into play as we believe others are paying more attention to our posts and lives than they actually are. All these cognitive biases shape the assumptions we make about others and ourselves.

The best strategy is to work hard not to make any assumptions that are bigger than, deeper than, or go beyond the specific content we are seeing in any given moment. For example, Theresa, an urban planner in New Mexico, described her approach to managing her assumptions when she's on social media: "I look at a picture, and if someone looks happy and like they're having fun, I allow myself to assume that they were actually happy and having fun at that moment. What I've really learned to work on is not carrying my assumptions further than that. I just assume that they felt that way right then. Doing that makes me feel lighter." I love this approach because it's simple and it's backed by evidence. Recent experimental work shows that actively using social media—like updating your page or searching for specific information—tends to reduce anxiety after a session. In contrast, passively scrolling and reading without a clear purpose is linked to feelings of anxiety. Another experimental study showed that practicing mindfulness techniques before logging on to social media helped people to absorb content holistically and critically and to jump to fewer conclusions, which reduced anxiety and feelings of loneliness and increased feelings of well-being. My takeaway is that if you're going to spend significant time on social media, be active and mindful so you can keep your assumptions confined to the specific post you see and not generalize them to more broadly represent the person who posted.

The most convincing finding, though, is that we should try to limit our use of social media if we want to avoid the exhausting act of assuming. In one experiment, participants took a one-week break from Facebook, Instagram, Twitter, and TikTok. These participants were matched to another sample with similar demographic characteristics who had no restrictions placed on their social media use. At the one-week follow-up, participants who stopped using all social media showed a significant reduction in ratings of depression and anxiety compared to their scores prior to the experiment. There were no differences in the group who continued their typical use of social media. In another experiment, participants were asked to reduce their smartphone usage to one hour a day for three weeks. Compared to the control group who had no restrictions placed on their smartphone use, the group who reduced their use showed significantly greater reductions in symptoms of depression, anxiety, and FOMO.

The implications are clear: Spend as little time on social media as possible if you don't want to be exhausted by all the assumptions you can't help but make. I have worked to limit the time I spend on social media by several hours a week over the last few years. Like you, I don't want to give it up completely because it does a lot of great things for me. But being on these apps less makes me feel less exhausted. And when I am on them, I'm active and diligent, and I try to channel Theresa and not assume that what I am seeing means anything more than what it does.

REMEMBER: IT'S NOT YOU, IT'S THEM

Reza frequently found himself irritated by his old friend Juan's social media posts across Facebook and LinkedIn. Whenever Juan shared vacation photos or announced a promotion, Reza couldn't help but assume he was either showing off or seeking validation. "I was always thinking the only reason I'd tell anyone that was if I wanted everyone to think I was cool or if I needed lots of external validation," Reza told me after describing one of Juan's posts that particularly irked him. The assumptions Reza was making about Juan's posts made him dislike Juan. "It's exhausting to see your friend act like an idiot, and it takes an emotional toll to realize your friendship isn't what it used to be," Reza said.

One day when Reza was driving home from work, he listened to a podcast that had a story about two pirates and a cheese sandwich. When he and I talked, he told me he couldn't remember the name of the person on the podcast or what the name of the theory was they talked about. But when Reza said the words "pirates" and "cheese sandwich," I knew immediately.

Rebecca Saxe is a professor of cognitive science at MIT known for her work on something called "theory of mind." The story she told went something like this:

> There is a pirate named Ivan who has a cheese sandwich. Ivan decides he needs a drink with his cheese sandwich, so he places it on top of a pirate chest and leaves. While he is away, the wind comes and blows the sandwich onto the grass. A second pirate, Joshua, comes along and puts his sandwich on the pirate chest. Just like Ivan, Joshua then goes away to get a drink. Then Ivan returns. Which sandwich will Ivan take?

Saxe's research shows that if you ask a three-year-old this question, he'll immediately say that Ivan will choose the sandwich on the ground. But if you ask a five-year-old, she'll say that Ivan will take the sandwich on the pirate chest. Saxe and her colleagues argue that between the ages of three and five humans develop theory of mind, which is the cognitive ability to understand that other people have different perspectives, beliefs, desires, and intentions from our own. A three-year-old who hasn't yet developed theory of mind can't recognize that Ivan experiences the world differently than he does. He can't process the fact that Ivan wasn't around to see the cheese sandwich blow onto the ground. Instead, the three-year-old believes Ivan knows everything he knows and, consequently, that Ivan will pick up the sandwich from the ground.

We need theory of mind to take other people's perspectives. By recognizing that someone else might see or interpret a situation differently, we can more accurately anticipate their reactions and respond with empathy and understanding. Perspective-taking involves using this understanding to consider situations from another person's viewpoint. It requires us to temporarily set aside our own assumptions to better grasp the other person's experiences and motivations. This is important because, as Saxe's research shows, theory of mind influences moral judgment. A three-year-old might decide that Ivan was being naughty for taking Joshua's sandwich because he didn't want to eat his own dirty sandwich and conclude that he should be punished.

"That podcast hit me hard," Reza told me, "and I decided I was going to try and not act like a three-year-old anymore." Instead of interpreting Juan's posts through the lens of his own motivations, Reza began to consider what might be driving Juan to share. He thought about what he knew about Juan and all the experiences they'd had together, and he slowly began to appreciate that Juan's vacation photos likely stemmed from a genuine love of travel and photography and his promotion announcements were rooted in a well-earned sense of pride. When he started to think about what Juan's motivations were, Reza's annoyance quickly began to recede. He saw that Juan's posts were expressions of his passions and accomplishments, not attempts to boast or seek attention. The empathy Reza developed through perspective-taking transformed his social media experience. As he told me, "I try to put this into practice all the time now. I don't just make assumptions about other people's motives based on what I would do. When I put myself in their shoes, I'm just a lot happier. It's like I made a choice not to go into some emotional war zone that'll wear me out." Actively working to take the perspective of others can help us sidestep cognitive biases that lead us to assumptions that will exhaust us.

A PERSON'S AI AGENT IS NOT A REFLECTION OF THEM

AI agents are just starting to enter our lives. An agent is an AI that can take in data, make a decision based on the input, and take action in response to that decision. Whereas ChatGPT and similar large language models can take in data you type or speak and give you responses, predictions, or recommendations, it doesn't decide what to do with its output and act autonomously. But AI agents do just that. With Camille Endacott, who is now a professor at the University of North Carolina at Charlotte, I studied an early AI agent named Lisa who was designed to help people schedule meetings. Lisa uses a conversational user interface (similar to ChatGPT), in which you could talk in natural language about what you wanted to happen. For example, if someone emailed you to ask for a meeting, you could respond to that person to tell her you'd meet and copy Lisa on your email and ask it to schedule. People we studied would write things like, "My assistant Lisa will find a time for us to meet in the next two weeks. Please work with her [yes, they most often said "her"!] to schedule." Lisa would input this prompt and interpret that you wanted a meeting with the person. It would then open your calendar and find your available time slots, make a decision about when you would like to meet and what priority you would give to this meeting, and then send a scheduling email to the person who requested the meeting. Lisa would decide when to send that scheduling email, what the message would say, what tone it would take, and with what frequency to follow up if the first volley didn't work out. If you think about it, it takes an incredible amount of decision-making, reasoning, and action to perform that sequence of tasks.

Lisa did it all pretty well. But as we all know, scheduling is tricky business. No matter how prepared you are, problems are going to arise. Lisa did its best to handle those problems, but it could only do so much. Inevitably, some person would get upset that Lisa didn't offer good meeting times, that the way it worded things was too confusing, that it came off as rude, or that it didn't respect the person's time well enough to respond in a timely manner. When that happened, the principal who deployed Lisa as their agent often had to intervene. For example, Richard was trying to schedule with someone he didn't know and who could potentially be a useful new connection. The person with whom Richard was trying to schedule received multiple emails from Lisa but never received a meeting time confirmation. Richard, who was copied on the email communications, could see that this person was getting mad. As Richard said, "I had to call him and apologize and say, 'Look, I'm really, really sorry. Lisa is not a real person, she's a bot, and I really apologize for the frustration.'" Camille and I expected that people would sometimes be frustrated and that the principal would have to intervene. But we were surprised to find that of all the people we talked to who had scheduled with Lisa, almost none got mad at it. Instead, they got mad at the person who made them schedule with Lisa. And they didn't get mad because they had to schedule with an AI agent, they got mad at the person as though Lisa were a representation of them. Another example illustrates this point.

Noah, who scheduled with Rahul by communicating with Lisa, was planning to meet Rahul at a coffee shop close to Rahul's office when he got an email from Lisa marked "urgent." Lisa asked, "Can you meet Rahul downtown in this bar?" Noah explained that Lisa kept changing the plan, which frustrated him: "I thought, 'What the hell. I'll get into a taxi.' Not very convenient. And then Rahul's assistant tells me as I'm in the taxi, 'Can you actually meet me here?' And I'm like, 'OK, fine.' So we go there, and just as a friend I say, 'By the way, mate, your personal assistant is awful.' They just wasted twenty dollars having me drive around town." When Rahul explained to Noah that Lisa was an AI agent, that only made the situation worse. Noah told us that he thought Rahul was rude and kind of a jerk. When we asked him if this was because Rahul used an AI agent, Noah said it wasn't but that it was obvious that Rahul didn't care enough to get the meeting right.

Noah and many others we talked to transferred their negative assumptions of the AI agent—rude, dumb, clumsy—to the principal. They told us about how they reflected on the negative interaction and got angry with the principal. But it wasn't the principal who was making any of the decisions; it was the AI agent. We also studied human executive assistants. They occasionally messed up scheduling too. But the annoyed people who suffered from these mistakes just got mad at the executive assistant. They didn't assume that the assistant's behavior was representative of the principal the way they did when they scheduled with Lisa. In general, the findings showed that communication partners transferred their perceptions of the AI agent's decisions to reflect on principals, for example, by assuming that principals were incompetent or impolite when the AI agent made questionable decisions. The only exceptions were when communication partners already had a strong relationship with the principal. In those cases, they typically compartmentalized their impressions to the AI agent rather than transferring them to principals. Other studies have shown similar effects. Of course, if your goal is to make a good impression with someone, maybe you shouldn't delegate the task of corresponding with them to your AI agent—a theme we'll return to in chapter 6.

Our changing communication environment is providing us with many more options to make assumptions that exhaust us. Let's remember that people are not the robots who act for them. Working to not make assumptions, even though it's so easy to do, is a clear path toward reducing our exhaustion.

Act with Intention

Marlon always makes a point of leaving his phone on the kitchen counter after work. He doesn't keep it in his pocket or bring it to the dining room table because he knows that if he does, he'll check it. But despite his best efforts to stay away from it and be present with his wife or play with his kids, he finds himself drawn to his phone. "It's like a tractor beam," he says. "It just pulls me in even though I'm trying to stay away." What makes Marlon feel exhausted and unhappy with this behavior is not that he can't stay away from his phone but that when he does pick it up, he's aimless. "Sometimes I tell myself, 'Oh, I need to make sure I didn't forget to send that email to one of our suppliers' so I have an excuse to look at my phone. But I know I didn't forget to send that email. I'm just lying to myself so I can see if anyone posted anything interesting on LinkedIn or check my fantasy football scores." Thirty-five minutes later Marlon puts his phone back on the counter. "I accomplished nothing. I just wasted half an hour of my life," he says.

Too often, we use our digital technologies without any sense of purpose. My own data suggest that even at work, people spend as much as seven hours a week opening programs, reading through communications, browsing files, and surfing the internet without a clear goal. All of that time spent adrift comes at the expense of other activities we could be doing. Tevfik, a regional sales manager for a scientific instruments company, estimates that his aimless use of digital technologies results in him spending four more hours a week in the office than he should be spending, and nearly one extra hour a day working after he gets home. As Tevfik says, "I'm exhausted from using all the digital devices, so I just sit there and kind of veg out. But then I look at the time I wasted and that makes me feel even more exhausted."

Marlon and Tevfik suffer a plight common to many digital technology users: They are drawn to their devices but lack intention in their use. They pick up their phone or sit down at their laptop aimlessly and for little reason other than that the device is there and beckoning. It's not so much the device as it is the potential of new information that draws us like a moth to a flame. Our apps are designed to stimulate the production of dopamine in our brains. Dopamine is a neurotransmitter, which is a chemical messenger that helps to transmit signals between neurons. It is often referred to as the "feel-good" neurotransmitter because it plays a key role in regulating mood, motivation, and reward. When we experience something pleasurable, such as eating a delicious meal or receiving a compliment, our brains release dopamine, which creates a sense of reward and pleasure. This reinforces the behavior that led to the pleasurable experience, making us more likely to repeat it in the future. We just keep wanting to come back to that app. Many times, we are on our phones to buy shoes or pay a bill, or on our tablets or laptops researching information for work, when we mindlessly shift to a game or a social media app that draws us into its infinite scroll or go down a search rabbit hole. These are all examples of how we proceed aimlessly through the world of digital content. As Tevfik told me, "I'm serious that hours can go by sometimes and I kind of wake up and think, 'Oh my God, what have I been doing for all this time?'"

A 2023 global study found that 41 percent of internet users aged sixteen to twenty-four reported that one of their top reasons for logging on to their devices was to "fill up spare time." And according to another study that focused on US teenagers, of the roughly 8 hours and 33 minutes teens report spending *actively* online each day, 2 hours and 47 minutes was spent simply browsing social media and websites with little purpose. If you add that to the 3 hours and 16 minutes that they report watching TV, movies, or video, which some might also say is without much purpose, that's more than half of the time they spend on digital technologies being sucked into a black hole of aimlessness. The statistics for the total time adults and teens spend simply browsing and otherwise frittering away their days is roughly equivalent in many other countries. The bottom line is that Marlon and Tevfik are more the rule than the exception. Like them, you, me, and our friends and family spend a lot of time lost in the ether.

The simple rule that we'll explore here is to act with intention when we reach for our devices. This rule applies equally when we're at work, out for dinner with our friends, or spending an evening at home. I am not suggesting that we limit how much we use our digital tools. Although, if you act with intention, you may end up using them less. The point of this rule is to make sure that each time we grab our phone or sit down at our computer, we're doing so with purpose.

THE PROGRESS PRINCIPLE

In one of my favorite business books, from which the title of this section is borrowed, Harvard University professor Teresa Amabile and her husband, Brandeis University professor Steven Kramer, present the results of a study examining the diary entries of 238 participants. The authors assessed what drove people to have meaningful and satisfying experiences at work, or what the authors call a rich "inner work life." Their conclusion: "Of all the positive events that influence inner work life, the single most powerful is progress in meaningful work." When we feel like we are moving forward and taking purposeful steps toward an important goal, we feel good. When we don't see progress, or worse, when we experience setbacks, we feel exhausted. A sense of progress is an energizer, and it keeps our exhaustion at bay. But what counts as progress, exactly? And how do you know that you made some?

In an answer to the first question, Amabile and Kramer discuss the importance of small wins. Small wins are bite-size accomplishments that one can reflect on and celebrate. They don't have to be major feats; success in seemingly minor activities can significantly impact one's inner work life. The daily diary entries submitted for their study were filled with reports of small wins that energized people: a scientist feeling joyful after someone said how good his experiment was, a programmer's excitement after fixing a nagging bug in her code. These small victories, although minor, accumulated to make a considerable difference in people's overall contentment in their work. Importantly, the experience of these small wins helped give people the energy boost to keep going, even though their entire project was not nearly complete and there were many hurdles left to go.

On the second point, the research suggests that recognizing progress in one's work requires awareness of actual improvements. Richard Hackman and Gregory Oldham, the godfathers of "job design theory," point to two primary pathways through which such awareness can be achieved. The first pathway involves feedback from a supervisor or knowledgeable peer—someone who affirms for us that

what we're doing is good, important, or useful. This external validation helps us to recognize and mark our achievements. Although acknowledgment of our progress from others is useful, Hackman and
Oldham suggest that there is an even better and more meaningful
pathway: obtaining feedback directly from the work itself. For in-
stance, a programmer who tests complex code and finds out that it
works as she planned receives immediate confirmation of progress.
This direct interaction with the work provides a sense of achievement
without needing external validation or contact with others. But a lot
of times, we don't design our work to give us direct and immediate
feedback on our progress. We might design the code, but then we give
it to someone else to test. Hackman and Oldham argue that a design
of jobs that splits execution from immediate reward is demotivating
and leads to exhaustion. In my own consulting work, I've used these
motivation potential scores to understand if jobs are designed to give
my clients enough feedback and to see what progress they've made.
I've seen time and time again that jobs that aren't designed around
these principles lead employees to feel demotivated and burned out.

No matter what activities people have to do at work or at home,
they report feeling less exhausted if they have a goal they are working
toward and a real sense that they are making meaningful progress toward that goal. But if we are going to have a goal and a way of measuring our progress toward it, we have to act with intention. Browsing the
internet for decorative socks, scrolling through social media to read
memes, or checking Slack just to see if anyone has messaged about
project updates are activities that do not follow the progress principle.
Our time feels wasted and we grow frustrated when we recognize that
we've gone in deep without purpose, when we could have been doing
something more.

One way to combat our counterproductive tendencies is to be

mindful when we use our digital technologies. Psychologist Ellen Langer, who has explored the concept of mindfulness in a variety of contexts, defines it as the process of active engagement in the present moment. When we are mindful, we notice what we're doing and are aware of why we're doing it. Mindfulness is essential for establishing goals and looking for feedback about whether we're making progress toward them. Without mindfulness, it's hard to recognize our small wins and it's easy to drift off track and into the abyss of aimlessness.

Several studies have found a direct link between mindfulness, progress, and exhaustion. They show that people who are mindful about setting goals for their digital technology use are significantly less likely to experience stress and exhaustion and feel a higher degree of progress in their work than users who do not act with intention. Additionally, people who are mindful enough to use their digital tools with intention for specific purposes are better at prioritizing their activities and focusing their attention, make fewer inferences about people and events, and feel in control of their actions. People who are less mindful in their use of digital tools report switching their attention often across domains and arenas and feel moderate to high levels of anxiety about wasting time while lost in mindless activity. These findings affirm the underlying tenets of the progress principle: If we act with intention when using our digital tools by setting goals for our use and making small wins toward them, we feel less exhausted and more energized.

Because they wanted to defeat their digital exhaustion, Marlon and Tevfik each changed their behavior to follow the progress principle in their own way. Marlon made sure that whenever he reached for his phone, he could articulate a real and important goal for his use as well as an indicator that would show what progress he made toward his goal. As he described, "If I decide I want to buy some running shoes, I'll tell myself, 'Your goal is find three good options.' Then, before I even pick up my phone, I'll come up with a game plan. Like, first I'll go to Runner's World and look up some reviews, then I'll go look at the Nike, Hoka, and Brooks sites. Then I'll go to Amazon and find the prices and put three in my cart." By specifying a goal and a metric to measure his success, and creating a plan to get there, Marlon made sure that he didn't pick up his phone for no reason. As he reflected, "I actually kind of like doing this. Now every time I look at my phone I think, 'Why? Why am I using it? And how will I know when to stop?' That gives me power and I feel good actually getting shit done."

Tevfik applied the progress principle at work. He told me that his biggest time suck was "going down a rabbit hole of research that didn't really matter. So much scrolling and clicking." By setting a goal and specifying small wins he not only knew when to stop but he was also able to feel good about accomplishing something: "The change I made using your rule," he told me, "is that I don't just open my browser anymore and start going from here to there and all over. I decide what kind of article I'm looking for and I decide where to go to look for it. I just think, OK, I don't need to browse forever, I just need to find, let's say, five studies that support our claims and then I'll be good. So I actually feel good when I get those five. Putting the number there helps me. I can achieve it, and I get a burst of positive feelings when I get there." Whether at home or at work, making sure that each intentional action with our devices culminates in some feeling of forward progress gives us an energy boost and stops the discharge that comes from the incessant switching, inference, and emotion associated with mindless motion.

CREATING A BUFFER

In Oscar Wilde's *The Picture of Dorian Gray*, Lord Henry, a friend of Dorian's painter, advises the eponymous protagonist that "the only way to get rid of a temptation is to yield to it. Resist it, and your soul grows sick with longing for the things it has forbidden to itself." Lord Henry was referring to a hedonistic lifestyle, but if Wilde were writing today, he might have just as easily been talking about smartphones or social media. Adam Alter, author of *Irresistible*, writes about how digital technologies are designed to tempt us. In talking about the lure of our devices, he notes, "Our willpower is limited, and eventually, as humans, we'll all fail in the face of temptation." Marlon admits he's weak when it comes to staying away from his devices. "Seriously," he told me, "probably the biggest temptation in my life is looking at my phone. I can resist just about anything else. But it's like I always give in to my phone. If I don't, I'm too antsy." I certainly know the feeling, though I'm not quite as tempted by my phone as Marlon. It does seem like my phone is a tractor beam, pulling me toward it even if there is no reason to use it. And it's not just because I have notifications turned on that draw me to the phone. The latest data suggest that notifications drive our interactions with smartphones in only 11 percent of all instances of use; 89 percent of our interactions with our phones happen because we pick them up and use them without prompting. So just turning off notifications, as many people suggest, is not likely to help much. Most of the time, I don't want to pick my phone up, but for some reason, I just feel like I need to. I can usually try to resist for a while but there's an unsettled feeling inside that doesn't dissipate until I yield to it by checking something that doesn't need to be checked. OK, maybe I do have it as bad as Marlon.

The loudest laugh I ever heard at an academic talk was when professor Jeff Hancock, who was then at Cornell University, came to Northwestern University to present his research about how use of our smartphones can literally affect the way our brains process stimuli.

He showed evidence from a study he conducted with a team of anes-
thesiologists about how patients undergoing minor surgery needed
fewer analgesics (drugs to control pain) when they were texting with
people during the surgery. No one laughed at that finding. But they
did laugh loudly when he suggested that the connection between our
phone and our brain is so powerful that most people in the audience
probably experienced "phantom phone vibration syndrome" during
his talk—they thought their phone was vibrating to signal an incoming call or text message when it actually was not. Then, as soon as the
laughing dissipated, just about everyone in the room pulled out their
phones to check.

Hancock's research in the operating room showed an exaggerated
effect that many of us experience in our kitchens and living rooms.
Chronic smartphone use changes our brain chemistry. Specifically,
heavy phone use disrupts the normal ratio of gamma-aminobutyric
acid (GABA) to other neurotransmitters. GABA is an inhibitory neurotransmitter that produces a calming or euphoric effect. Having too
much GABA can result in a number of side effects, including exhaustion and anxiety. Drugs like heroin cause an increase in GABA activity, which depresses the central nervous system. The research seems
to suggest that chronic phone use can lead to similar changes in our
brain that mirror the effects of illicit substances like heroin. Anna
Lembke, who runs the Addiction Medicine Dual Diagnosis Clinic at
Stanford Medicine, calls smartphones "the modern-day hypodermic
needle." All this is to say that when we feel like we have trouble resist-
ing the urge to hop on our phones, there may be a physiological driver.
As Lord Henry understood, our brains make it feel good to give into
the temptation.

That physiological driver is also reinforced by the power of habit.
Habits are largely the result of unconscious action. When something

reaches the threshold of a habit, we typically stop thinking about what we're doing and just do it. If you're like me, Marlon, or Tevfik, you have many habits with your digital technologies, some of which you recognize upon reflection, and some of which you don't. For example, sitting down at your computer and immediately opening your web browser is a habit. So is pulling your phone out of your pocket several times a day just to look at it. As Tevfik told me, "I just open Pandora all the time. Not sure why. Most of the time I'm not planning on listening to music. My thumb seems like it's attracted to the app." Apart from whether our smartphones and applications change our brain chemistry, we certainly use them in habitual ways without thinking. Studies have shown that reaching for smartphones, checking Facebook, sending emails, and even turning to ChatGPT all quickly become habits that people have a hard time breaking. Whether they need to grab their devices or use those apps is not something they consciously think about; the patterns of use become so habituated that they are difficult to stop. Now, add to those habits changes in our brain chemistry that might be occurring as both the result of and driver of our habits, and it's easy to see why Marlon, Tevfik, and the rest of us have a hard time staying away from our digital tools, even though we might want to.

It's hard to act with intention if you're not thinking about your actions. That's why most of the people I've worked with who have been successful in adopting this rule have figured out ways to create a buffer between themselves and their digital tools. Just adding obstacles between yourself and the things you're automatically drawn to can create enough of a barrier to break you out of your habits. James Clear, the habit maven and author of the masterpiece *Atomic Habits*, provides a compelling example. He talks about how he would often open his refrigerator, see a can of beer in the door, and automatically grab it. Simply placing the beer in the back of the fridge where it was more difficult to grab broke the automatic behavior of reaching for it. Now that there were obstacles between him and the beer, he had to think (even if for a split second) about whether he wanted the beer enough to move stuff out of the way to get it. That small hesitation might not be enough to stop someone who really wants to drink a beer or has an addiction to it, but it is often enough to break the automaticity of our behavior. Much of the time, we don't really want to check our phone or open our web browser, but there are no obstacles in our way to cause us to question our behavior. Obstacles allow us, as Meryl Reis Louis and Bob Sutton observe, to "switch our cognitive gears" from habits of mind to active thinking.

So how do we actually switch our cognitive gears when reaching for our digital devices and start acting with intention? I collected a list of helpful practices from people I've interviewed over the years who have successfully created a buffer between themselves and their digital devices. Perhaps they'll spark your imagination as you create your own buffers.

Ideas for Creating Buffers Between Yourself and Your Digital Devices

IDEA	WHY IT WORKS
1. Buy a charging station for devices like phones and laptops and put it in a central place for everyone in the family to see.	"When all of the phones and tablets are there, you can see them and it makes you think, 'Do I want to be the only one to take mine out?' Also, my wife and kids will know that I'm on my phone if they see it missing and then I better have a good excuse for why." −MARLON

IDEA	WHY IT WORKS
2. Designate a texting hour.	"For me, I just plan on texting back everyone who's texted me in the evening around 8:30. I check my phone then and fire off texts. Knowing that I've got that time to check makes me feel less like I need to constantly look to see if any text has come in." –TEVFIK
3. Turn off notifications on your devices.	"If I see my phone light up because there is a notification I'm literally like a moth to a flame. But if I turn them off, my phone doesn't light up. Sounds simple but it works. I just don't look at my phone as much. And none of those notifications are ever about anything urgent." –RENEE
4. Put your apps into folders.	"I put all of my apps into folders. So when I look at my phone I can't mindlessly click on any app or see an immediate notification. I have to remember what app is in what folder and a lot of times that's enough for me just not to use it." –KATHRYN
5. Disable fingerprint/ face recognition.	"I know it's weird, but having to type in my passcode to open my phone is all it takes to remind me to ask myself 'What are you doing on your phone?' and 'Do you really need to be doing it?' I didn't have that barrier when the phone automatically unlocked for me." –BART
6. Keep your laptop at your desk plugged in to an external monitor.	"I used to have my laptop close by and I'd open it all the time. Now, I keep it at my desk plugged in to my monitor. If I want to use it, I go to my desk and have to sit down and that feels like I'm at work. So then I ask myself if I really want to be at work instead of watching a movie with the family." –MIKE

IDEA	WHY IT WORKS
7. Have just one Netflix and Spotify account for the whole family so your recommendations are mixed up.	"I switched to having one account for the whole family for my streaming services. That way I get recommendations for bad movies and songs I don't want—stuff my kids would like—and they get stuff that the algorithm thinks I want because it can't tell who's who. That means that now I know that if I log in to one of those accounts, I'm going to get a bunch of crap to wade through and I'll be annoyed. So a lot of times I just read a book instead." –ÁNGEL
8. Disable tabs on your web browser so you have to consciously click a link or type in a new URL to get to new content.	"If I have lots of tabs open, I go back to them and look. Then I click on a link on them and spiral out of control. So I just disable my tabs so I can only have one open at a time. Then I have to choose if I want to go to the next thing or not. Most of the time that stops me from doing it." –XIN
9. Tell people to call you. Let them know you won't check text, email, or socials after work but are available by phone.	"I've gone old-school. I just tell people if they want me after work—work people or friends or whomever—they need to call me. I won't email or text. That way they have to make the choice if they really need me enough to call, since no one calls anymore. That works for me, because I can put my phone somewhere else that won't tempt me and I can hear it ring all over the house" –BRUCE
10. Get rid of shortcuts on your devices.	"If something is there, you're going to just click on it and not even think about it. I've found [that] to break that habit I get rid of things to click on. I've gotten rid of all my shortcuts. If I want to go somewhere online or get to some app, I have to actually go there now, which is just enough work to make me pause for a second and think about it, so I don't just do it out of habit." –SOFIA

Hoping for some fun and creative answers, I also asked several generative AI tools what they might recommend. I prompted them as follows: "Give me ten unconventional ideas of things a person can do to create a buffer between themself and their digital devices so they don't just use them out of habit. For each idea, explain why it will work." I got some boring answers and some that were very wacky. ChatGPT told me to hire a "Phone Guard" to physically take my phone away for designated periods. It explained that "social account-ability is powerful. Knowing that someone else controls your phone access adds a layer of responsibility and inconvenience, making you less likely to ask for it back unless necessary." And Claude, built by Anthropic, told me to wear oven mitts around the house. As it re-minded me, "The bulkiness and awkwardness of oven mitts make using a touchscreen or typing at a keyboard nearly impossible without removing them, adding a significant inconvenience to phone use."

Some of these suggestions aren't as crazy as they first seem. In 2024, a company called FTLO Travel began offering phone-free trips

"It keeps me from looking at my phone every two seconds."

to locations around the world. Clients who book must surrender their phones upon arrival and don't get them back until the last day. "Removing that sort of temptation has always helped facilitate better bonding and conversation," says the company's founder. The company had bookings of nearly three thousand travelers for its phone-free excursions in the first year it offered them. At one hotel on the Mexican Riviera, guests can opt for a "detox concierge" who will remove the hotel room's TV and lock all of the guest's personal electronic devices in a safe.

As Marlon reminded me the last time I talked with him, "I actually like using my phone to do things and learn things. But I do feel a lot less exhausted if I go on knowing what I'm doing and what I hope to get out of it. I guess that's the point of being intentional, right? That makes me feel much more energized." That is, after all, the goal. But sometimes to act with intention we need to be reminded to form intentions. Erecting barriers that create just enough friction to get us to think about why we are using our devices, when we'll know that it's time to stop using them, and what would make us feel successful when we stop, is the heart of this simple rule.

Learn Vicariously

The first six rules teach us how to use our devices in ways that stop them from exhausting us. This rule and the next show us that we can use our devices in ways that provide us with new sources of energy. An example from a project I worked on with Discover Financial Services—a US bank with more than seventeen thousand employees—shows how. When I first met employees in the marketing division at Discover, nearly everyone mentioned how exhausted they were and how using so many different digital tools was draining them. From junior assistants to senior vice presidents, people lamented that all the time they spent working with data and communicating with one another was getting to feel like too much.

Just as feelings of exhaustion were peaking, management decided to add another tool to the mix: a new social networking platform built by a company called Jive. Think of Jive like Facebook for your company. The platform was limited to Discover employees, who could post and connect with one another in the same way they did on platforms like Facebook. The company implemented the social networking platform to provide yet another way employees could connect and share data and knowledge with one another. As you might imagine, employees were less than thrilled. Not only were they going to have to learn how to use another digital technology, the new tool would be another thing they'd have to check and another conduit for information they'd have to deal with.

Because morale was so low, I talked the senior leaders into letting me run an experiment with their new social networking platform in two of the company's divisions. The first group of employees was in the marketing division. Marketing had ten departments and several hundred employees. The marketing division would have access to the Jive tool, while the other group—the operations division—would not. Operations had nine departments and roughly the same number of employees with the same demographics, so they made a good comparison to marketing. Before the study started, I administered a battery of surveys to both groups, looking to identify how accurate employees were at identifying experts—people they could go to if they had questions about a particular issue. I also tested how well they could find people who could make connections for them to employees who had knowledge or resources they needed. After these initial tests, the marketing group used Jive for six months and the operations group did not. Then I administered the same survey that we gave before the study started.

I was amazed by the results—and so were the company leaders. The employees in marketing who used the social networking platform improved their ability to find experts by 31 percent. They also improved their ability to accurately identify who could put them in contact with experts by a whopping 88 percent. The operations group—the group that did not use Jive—showed no improvement on either measure over the same period. As we probed deeper, we learned why. The marketing group was exposed to communications among employees via Jive wall posts, direct messages, and likes. By seeing what their coworkers talked to one another about and with whom they interacted, employees were able to learn what and whom those coworkers knew. The operations group was not exposed to these ambient communications and, consequently, did not increase their learning about what and whom their coworkers knew.

One of the major benefits of using digital technologies that is hard to replicate by any other means is the access they give us to new and different sources of knowledge. We can take advantage of this access to others to learn vicariously—that is, by simply watching others without the need to interact with them directly. When we expand our ability to learn vicariously, we discover new things about people. We can innovate better, solve problems faster, come to see the world differently, and make deeper connections with others. A long-standing line of research I've conducted shows that all of those things give us new energy. The problem is that we don't often know how to unlock the power of our digital tools to provide us with those benefits and "see around corners," as one of the marketing execs at Discover put it.

ON THE BENEFITS OF EAVESDROPPING

In today's knowledge-intensive organizations, much of the work that people do is hidden from view. Employees sit at their computers, typing reports, performing analyses, writing copy, and carrying out tasks that are not easily observable by others. Our postindustrial economy increasingly disembodies work, leaving few physical manifestations for others to witness. This invisibility is exacerbated as organizations break down work into smaller units, distributing it across various teams and departments that are often spread out geographically. Even when work does have a physical element, the dispersion of tasks means few colleagues are around to see it. Consequently, Bonnie Nardi and Yrjö Engeström, two leading scholars of changing work dynamics, observe that in a knowledge economy, "Work is, in a sense, always invisible to everyone but its own practitioners." We can work right across the hall from someone for years, or live right down the street, and still be unaware of what each other knows.

At Discover, just like most other companies, employees generally had a clear understanding of what their immediate team members and nearby colleagues did every day. However, they often had little insight into the activities of coworkers from different departments or those located in other offices. That work was invisible to them. The main reason for this knowledge gap was a lack of communication. Employees often learned about their coworkers' work by sitting in meetings with them, being staffed on a team with them, or receiving some report or other deliverable from them. In other words, they learned by direct experience with them. Ginny, a senior associate in the marketing group, described how she typically learned about others' work: "I typically interact with people I know something about—know what they do. That mostly means I communicate with people on my immediate team . . . I don't ever really see those people actually doing work. I mean, I see them sitting at their computers, but I don't know what they're doing. But I normally get a sense of what they do because they tell me about it when we talk, or I get copied on an email they send to someone else and I read what they say, or I overhear them talking about some aspect of their project with someone who comes to their desk." Ginny's description of her interaction with work colleagues at Discover could just as easily describe a thousand other companies. We normally learn what others know as we interact directly with them. But the problem in most of our contemporary organizations is that we interact with the same people all the time. The research shows that our workplace networks are not very diverse, and that we tend to have repeat experiences with people who are on our teams, in our department, or who sit physically close to us. That means we see a lot of the same stuff over and over again.

Social media and other digital tools that allow us access to people from different parts of our companies—if we're talking about corporate platforms like Jive or Chatter—or from different walks of life—if we're talking about public platforms like Instagram or LinkedIn—provide a different opportunity for learning what people know. They allow us to learn vicariously. There's a long line of research that shows that people can learn effectively by watching. In some manual jobs, like those performed by machine operators or midwives, people often start their work as apprentices by watching experts for extended periods, after which they can typically perform the hands-on work with a high level of dexterity. In knowledge work contexts, people don't often learn how to do a job primarily by watching someone. But watching them or listening to them talk about their work are prime ways that we can develop accurate knowledge about what those people know. Knowing what and who other people know is what social psychologists call "metaknowledge," a fancy word that means knowledge about knowledge. There's lots of work showing that accurate metaknowledge is associated with better work productivity, fewer errors, and overall satisfaction with one's job. It's helpful to know what people know and who they know, but it is also invigorating to learn about new concepts and ideas and discover new people and new knowledge. As Jamon, an electrical engineer at a medical device company I worked with once, told me, "The most exciting thing in my job is learning what new things there are to know and figuring out who knows them and who could teach me. If I'm ever feeling bored or just kind of burned out, finding out who knows what gets me energized and excited."

On social platforms, we are third parties to goings-on that don't involve us directly but to which we have a front-row seat. And as my research with Discover showed, the way our social platforms allow us to eavesdrop on people who we wouldn't normally come into contact with or see regularly can improve our metaknowledge. Regina, on one of the marketing teams, provided a great example of what that eavesdropping looked like. As she told me, "I saw some messages exchanged between two guys in another department about how they determined the appropriate rate for a consultant. I didn't know how to do that and I thought it would be good to know, so I kind of made a mental note that these guys knew that so I could go ask them in the future. That would really be helpful knowledge to have for some upcoming projects . . . If they sent that message through email, I wouldn't have ever known that message was sent and I couldn't have seen it, so I wouldn't have learned that these guys know about rate determination." Regina had a front-row seat to a boring work conversation between two colleagues that was rich in information. To the two guys interacting directly, this was just everyday talk. But it was filled with useful information that Regina used to update her metaknowledge and build out her understanding of who knew what and who knew whom in different departments.

These improvements in metaknowledge were themselves energizing. But they were made much more so when they translated into new ideas or resulted in some form of innovation.

For example, Marta, who worked in the cardmember marketing department, had spent several months conducting research on why consumers were likely to choose one credit card brand over another. Her research uncovered that, within the demographic she was interested in, consumers made decisions largely based on the availability of rewards programs and, more specifically, rewards that could be redeemed for cash or credit on a card statement. Marta tried unsuccessfully for several weeks to figure out a strategic plan based on her findings. One day, however, Marta had a breakthrough. She recalled seeing a communication on Jive between two colleagues. One of them was a member of the pricing and analytics department. He had mentioned something about variations in interest rates being influenced by spending habits. This piqued Marta's interest, and after reviewing the conversation, she realized that there must be some flexibility in how rates were assigned. Realizing the potential, she suggested implementing a cashback bonus for consumers in a particular spending category, ensuring it stayed within their rate assignment limits. She reached out via email to one of the guys from the post to get more details, found the concept feasible, and built a program around it. "It's been really successful so far," Marta told me. "It was really innovative, so it made me proud." Marta produced this innovation by recombining knowledge in new ways. Importantly, she was enabled to do so by her awareness of what a stranger-to-her colleague knew because she had visibility into his communication through the social networking platform. People from across marketing discussed numerous other cases similar to Marta's, most of which were much smaller in scope and significance, in which they were able to combine knowledge that already existed within Discover in new ways because Jive allowed them to see what knowledge other people had.

I've since worked with five other companies who have used social networking and digital collaboration platforms to "see around corners." In each of them, learning vicariously led to ideas that energized. But that path was not always straightforward, and not all people reaped these benefits. Next, let's look at what kinds of changes we need to make as we use our digital tools to turn our eavesdropping into an energy boost.

THINKING DIFFERENTLY ABOUT YOUR CONNECTIONS

The first lesson I've learned about how to unlock the energy-enhancing power of eavesdropping through social tools is that we have to think about our networks differently. Most of us have a lot of connections with people on our social media and digital collaboration tools. But of those many connections, we only pay attention to and have meaningful relationships with a small few. Robin Dunbar, an anthropologist at the University of Oxford, proposed that humans can really only maintain about 150 meaningful connections. This is popularly known as "Dunbar's number." Dunbar argues that there is a correlation between the size of the neocortex—that part of the brain involved in higher-order functioning such as sensory perception, cognition, and language—and the maximum size of a social group that an individual can maintain. His research observed that primates with larger neocortices tend to live in larger social groups, and he extrapolated this finding to humans, suggesting that the size of the human neocortex limits the number of relationships we can functionally process. The idea is that the cognitive load required to manage complex social interactions and relationships increases with the size of the group, and the neocortex has a finite capacity for handling this complexity. Whether you agree with the number 150 or not, the point is well taken: We simply can't pay attention to all our work colleagues on Jive or all our friends on Facebook. So what do we do? We circumscribe our networks by focusing on a small subset of contacts—reading their posts, looking at their pictures, and responding to their comments—while mostly ignoring the rest. You might argue, "Not me. I scroll through and read stuff from lots of people I don't know that well or with whom I don't regularly talk offline." That's what most people say, including me. The data show otherwise.

A study published in the journal *Nature* reported on data analysis from the entire population of active adult Facebook users in the US in 2020. The data showed that "content from 'like-minded' sources constitutes the majority of what people see on the platform." The authors argue that this is likely due partially to people's own choices about who they listen to—people pay attention to others they think are similar to them and they think are relevant—and also partially due to the algorithm's selection, as it shows us content from people we have things in common with and who we are likely to perceive as similar to us or others we know. Another study conducted by researchers at the University of Washington and Microsoft based on data from users of Microsoft's digital communication tools showed that the number and diversity of connections declined significantly during the pandemic as people began to work from home en masse. People in knowledge-intensive jobs increased their interaction with people in their work groups and departments and dramatically decreased active interaction with as well as the amount of attention they paid to people outside their departments. The networks we cultivate on our digital tools might be large, and the opportunities for vicarious learning from them may be great, but in actuality we pay attention to a very small part of our networks.

With Luke Rhee, a professor at the University of California, Irvine, I conducted two studies with Korean software companies that illustrate how our orientation to our networks limits or creates possibilities for us to learn vicariously. We were able to track the people with whom engineers in various product-development functions maintained contact through their digital tools and in face-to-face contexts over an entire year. We were also able to track what ideas they submitted for process and product improvement in their organizations and how innovative their senior managers ranked those ideas to be. In line with the studies above, we found that of the many connections people developed and maintained through their digital tools, they actively paid attention to only a small subset. The people they tended to pay attention to were those with whom they worked on projects. We found that they did pay attention to a number of other people at the company whom they didn't work with regularly, but only if they suspected that those people might have information that was relevant to their work. We called this effect an "attention bias": we pay more attention to people who we think will give us useful data or information. But our findings showed that the engineers who were most likely to come up with highly rated innovations were those who were able to overcome their attention bias. When they paid attention to people they didn't work with or didn't think had much useful information for them, they came up with much better ideas (and we found they got better raises and were promoted faster for it). Why? Because most innovation comes from making connections among things that aren't yet connected. Innovation isn't about making something new that has never been seen before; it's about finding new problems for existing solutions or combining ideas in new ways. But to make those connections, people need to pay attention to parts of their network that don't seem obviously fruitful or relevant and learn vicariously from them.

The people at Discover who succeeded in energizing themselves through their use of the social networking platform did so by deliberately expanding their networks and making a conscious choice to cultivate and maintain a group of contacts different from those with whom they regularly communicated as part of their job. A manager named Jordan in the credit card marketing division discussed how such an expansion was a deliberate strategy: "We're a company of, what, fifteen thousand people? In a day, I get to interact with maybe seven of those people and in a week maybe four or five more. And I only work with four of them on my team. So Jive is an opportunity to get to know other people and interact with them and have experiences with them. So I don't want to waste my time being friends with people on there that I talk with anyway." As Jordan was aware, use of the social networking platform had the potential to put him into contact with people at the company who he would otherwise have little occasion to meet and with whom he would have few opportunities to build and maintain a relationship. The nature of these relationships on the platform was casual; they were not often strong. But the cultivation of strong ties was not what Jordan and others like him had in mind. Instead, they recognized that weak relationships that took little effort to maintain were quite desirable. As Bekah, an accounts specialist, told me, "On social media platforms inside the workplace or outside of it, you can have relationships with people without having to do a lot. You watch what they say to their teammates, you offer a comment every once in a while, and you just sort of get to know about them and what they do. But you don't have to take them to coffee or buy their kids a present. It's easy. And really, all I want when I'm on there is to get a sense for what people are doing; I'm not looking for a new best friend." For employees at Discover and in the other companies I've worked with, deliberate expansion of networks and attention to seemingly irrelevant others seemed, for most, to outweigh the limited social costs needed to maintain the relationship and the focus. Having access and paying attention to people with whom they otherwise would have little occasion to interact means that an employee can expand their network and share experiences with people outside their own work team in ways that lead to building better metaknowledge and producing more innovative ideas.

Although these examples are from inside companies—because that's where I've done most of my research on this topic—their import certainly doesn't stop there. As Brenda told me about her use of Twitter and Instagram, "I get energized when I learn new things about people or new ideas from outside my circle. I try to really only interact with people on my socials that I wouldn't see in normal life, 'cause that's where I learn the most and that is way more powerful for me." Bonnie Nardi's anthropological study of the video game *World of Warcraft* provided similar examples. She recounted how players would form teams and alliances with others on their quests that resembled "lightweight" relationships that didn't take much effort to maintain but allowed her study participants to learn from people they didn't otherwise know. These interactions were energizing in many ways, not least because exposure to new groups brought participants new skills and ideas about how to play the game better. Broadening our networks is work. But it does not take a lot of work to maintain connections to those whose experience we draw on vicariously. And doing so can be quite energizing.

THINKING DIFFERENTLY ABOUT YOUR THINKING

A second lesson I've learned about how to turn eavesdropping on our digital tools into an energy-producing activity is that we have to rethink how we think about and put together data.

My favorite description of the way we experience data through social media and most other digital collaboration tools we use comes from the journalist Clive Thompson. He writes (about Facebook specifically) that "each little update—each individual bit of social information—is insignificant on its own, even supremely mundane. But taken together, over time, the little snippets coalesce into a surprisingly sophisticated portrait of your friends' and family members' lives, like thousands of dots making a pointillist painting. This was never before possible, because in the real world, no friend would *bother* to call you up and detail the sandwiches she was eating." As Thompson points out, one pic or one post about a sandwich is meaningless, perhaps even annoying. But if we see enough posts about someone's sandwiches, we can learn something about the person eating them, like that they're always hungry at 2 p.m., that they don't like mayonnaise, that they're a vegetarian, or that their apartment is two blocks from a great sandwich shop. The bits and snippets can add up to more if we know how to look.

At Discover, the people in the marketing group that used the social platform oriented to this new data environment by changing their
behavior. Vince, a more senior employee, likened learning on social
media platforms to being a student. As he described, "If you ask the
teacher something directly, you know how the answer fits into the
bigger picture because you asked about it. If you overhear the teacher
talking to another student, you can still learn a lot, but you've got to
figure out how to contextualize it." On social media platforms, Vince
says that he can see what "a bunch of other people that I wouldn't normally talk to are saying to one another and all that conversation has
useful tidbits in it." To make that information useful, he has to change
his behavior. As he described, "I've got to make it a process to read
those communications or at least look at them quickly, and I have to
reflect on them a bit more than I would if I talked to either of those
people directly because I've got to try to pick up the context of the
comments." That contextualization is important. It often requires developing hunches about a person's knowledge and then looking for
information that confirms or denies those hunches. In the same way
that you might figure out that someone is a vegetarian by scrolling
back through all of their sandwich posts to see if they ever ordered
turkey or ham, diving through people's posts about specific jobs they
did or conversations about a particular task can reveal broader patterns. The key is knowing where and how to look.

The typical way people deal with problems they encounter in their
work or life is to reactively search for knowledge that can help solve
them. The water in your shower just won't get hot, so you google potential causes. Or you can't figure out how to run a specific statistical
analysis, so you go to a textbook to look for information about multi-
variate techniques. We might characterize this process as reactive
because people don't think about acquiring new knowledge, or the

metaknowledge that might help them get that knowledge, until the problem arises. We would characterize their behavior as searching because they are actively looking in a focused and directed way for some specific type of knowledge.

But the employees in the marketing group who were able to learn vicariously did not reactively search for knowledge when they ran into a particular problem. Instead, they proactively aggregated the metaknowledge they were acquiring each day as they spent time on the social networking platform. In other words, they stumbled into knowledge of "who knows what" or "who knows whom" at random intervals and, although they didn't have a particular use for that metaknowledge at the moment, they held it in passive memory for use at a later time. As Deena told me, this subtle change was actually quite profound: "I find myself doing something new now when I'm on social media. I see all these things that people are saying and I get a sense for what they know. So when I see something and realize that John knows about consumer promotion rates I sort of take a breath and tell myself that I should remember that, and I file it away. That's a pretty big change—to see something and try to absorb it so you can use it later when you don't even know if you'll ever need to use it." This shift in behavior paid large dividends.

When faced with a problem, these successful social media users recalled bits of information they'd stored and then went through a complex mental process of connecting these pieces to form a complete solution. Creating these comprehensive pictures of who knows what requires significant mental agility. Key cognitive skills like collecting, abstracting, and filtering information are essential for this integration. If social media and other digital tools are to be effective in helping us to learn vicariously, we need to develop the cognitive skills necessary for abstract thinking and linking various pieces of information. Although doing these feats of mental gymnastics might sound exhausting, and surely they require some cognitive exertion, my research shows pretty clearly that any tiring effects are more than offset by the thrill associated with learning and the uplifting energy received by presenting new ideas that make things better.

Be Here, Not Elsewhere

Mariana is a professor at a southwestern university. She teaches and researches about education and social policy reforms to promote diversity in schools. A review of her calendar for two weeks showed that she spent sixteen hours responding to messages on email and her research group's Slack channels, twelve hours in meetings (mostly on Zoom), six hours doing data analysis, eight hours writing a research paper, and six hours teaching graduate students over Zoom. During all these activities, she was actively engaged with one or more digital technologies. Her calendar also shows a curious set of entries called "tiling." Those entries occur early in the morning and in the late afternoon on several days, as well as for nearly six hours on Friday of the second week. When I asked her what "tiling" was all about, she responded proudly: "I'm tiling my bathroom—the floor and the shower. Oh, and I'm doing a super cute backsplash behind the sink." Mariana reported a 2 on the digital exhaustion scale.

Most people who have a similar mix of activities on their digital devices report much higher levels of digital exhaustion than Mariana. They complain that they spend too much time in front of a screen, that they have too varied a list of activities across too many modalities, and they feel guilty doing non-work activities like tiling the bathroom because their exhaustion already causes them to be behind on work and they feel guilty having fun. Moreover, they can't enjoy their extra-curriculars because their head is still back in Zoom or in emails they should have sent.

So how does Mariana do it? How can she pack her schedule with work and non-work activities and still report such low exhaustion? "I'm no superwoman," she told me. "I don't have an ability to take on more than other people. My kids actually joke that I'm a weakling be-cause I can't really do two things at once without messing up or feel-ing overwhelmed." Mariana's secret is that she has mastered the art of being present in each activity she engages in, without thinking about, worrying about, or getting excited for other activities she's not doing at the moment. She's here, not elsewhere. Mariana credits her ability to "be here" to a concept called "flow," popularized by the psycholo-gist Mihaly Csikszentmihalyi. Flow is characterized by intense focus, a sense of timelessness, and a feeling of being completely absorbed in the present moment. Typically, flow occurs when the challenges of a task are balanced with an individual's skills and abilities. It typically arises in activities that are challenging enough to require effort and concentration but not so difficult as to become overwhelming. In a state of flow, research shows that individuals experience a sense of deep enjoyment and fulfillment. They feel energized. They want to be where they are, not elsewhere. But most people don't experience flow when using digital tools because we divide our attention, make infer-ences, and let our emotions get the best of us.

People like Mariana follow this last simple rule: they make sure that they are present and engaged in whatever activity they're doing at the moment—whether on digital tools or not—and they make sure that their mind is there, not elsewhere. But they also do a second thing that is so important. They make sure that their portfolio of activities changes in ways that allow them to detach from what they did before, and they pick new activities that exercise distinct, but complementary, brain and physical muscles. This mix of immersion in work when using our digital technologies and strategic detachment from them is the secret sauce that powers this simple rule.

HOW TO STOP TELEPORTING

Dana #1 was six months into her first job as an assistant account executive at a PR firm. She was an English major in college and loved the idea of working in the firm's health care practice to secure media placements for her clients. "I always wanted to work in PR" she told me. "But I seriously didn't think how exhausted I'd be." Dana felt overwhelmed by all the different databases, tools, and communication channels available to her at work. Her clients all had preferred ways of communicating with her as well, and those often were on different digital tools than the ones she used with her colleagues at work. Over a period of months, Dana reduced her digital exhaustion by following several of the rules we've already discussed. She got rid of half her tools, worked on matching, reduced her assumptions, and was doing much better at acting with intention. But she still felt somewhat exhausted. After doing some serious thinking about her work and life, she came to a realization: "I honestly think I'm kind of bored. I'm reading through articles, reading bios on editors, finding media mentions of our clients and stuff like that. But it's not really all that challenging." She continued, "So I just find myself opening TikTok on my phone or googling random stuff while I'm supposed to be working. You know, all the stuff I'm not supposed to do. But I can't help it, and I'm still feeling tired."

Dana #2 was nearly thirty years into her job at the seventh company of her career. She was senior council for ethics and compliance at a large medical device company. She also loved her work. "Lawyers get a bad rap. But I really feel that what we do helps make sure we have a great company that is a great place to work," she said with pride. Dana #2 also felt overwhelmed by all the different digital tools at work. I met her because her team was implementing a new digital compliance tool to help the company meet the requirements of the Sunshine Act, which was part of the Affordable Care Act, a signature piece of legislation from the Obama administration that required manufacturers of drugs and medical devices that participated in US federal health care programs to report certain payments and items of value given to physicians. Upon reflection on her own work, she confessed, "I honestly think I'm just overworked and stressed. There is too much going on and a lot of the time I find myself just sort of mentally teleporting to a different place because my brain needs rest. Like, I come out of a daze thinking about a trip I took a couple of years ago or something when I'm supposed to be writing an important memo. That happens a lot these days. I'm just not present, and that makes me feel exhausted."

The situations that each Dana found herself in are well described by the Yerkes-Dodson law. Shortly after the turn of the twentieth century, psychologists Robert Yerkes and John Dodson conducted a series of experiments in which they placed mice in mazes. At the end of some paths were food rewards, and at the end of others were electric shocks. They varied the intensity of the shocks to create different levels of arousal in the mice and then measured how quickly they learned to navigate the maze to find the reward. They found that when the shock was mild, the mice took a long time to learn the correct path. When the shock was high, they also took a long time. But when the shock stimulated a moderate level of arousal in the mice, they were fastest at learning to navigate the maze. Yerkes and Dodson concluded that there is an optimal level of arousal for peak performance. Too little arousal and a mouse wouldn't work hard. Too much arousal and they would work too hard and burn out. But at just the right level of arousal, they were at peak performance. Extrapolating from mice to humans, as early psychological researchers were wont to do, they suggested that just the right amount of stimulation and stress keeps us engaged, focused, and at the top of our game.

Dana #1 sat somewhere on the left side of the Yerkes-Dodson curve. Her work wasn't demanding enough to sustain her attention, and her mind teleported her to other places out of boredom. Where her mind went, her taps, swipes, scrolls, and typing went too, leading her back into the trap of digital exhaustion. Dana #2 sat at the right of the curve.

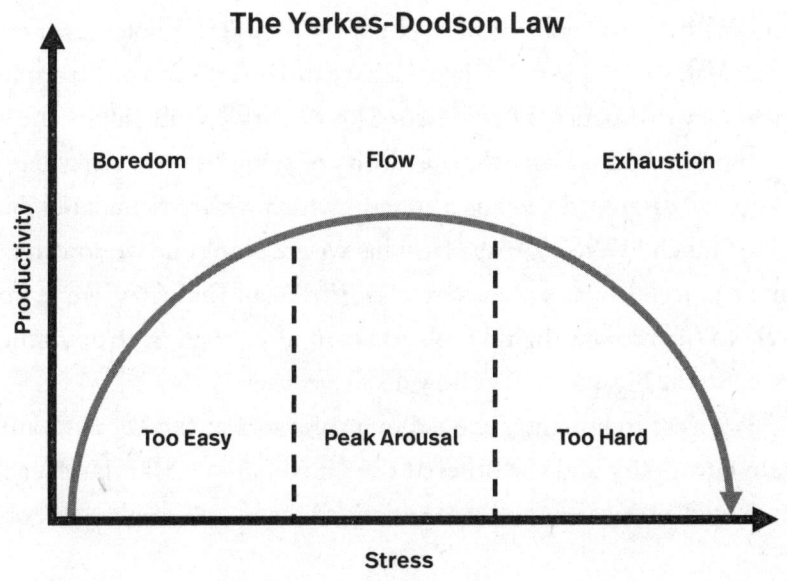

Her work was too demanding, and she had difficulty keeping the intensity of her focus. Research shows that when we sustain our focus for too long and engage in cognitively demanding tasks, we experience cognitive fatigue. This happens due to buildup of something called glutamate in the brain's prefrontal cortex. Glutamate is an amino acid and the main molecule that neurons use to signal one another. This buildup can make it harder to perform other prefrontal cortex activities, like decision-making and planning. When glutamate builds up, people often find themselves choosing actions that require little effort but offer moderate reward—like scrolling through LinkedIn or looking at pictures of a favorite vacation in their photos app. In short, the cognitive fatigue we experience at high levels of arousal leads us to use our digital technologies to mentally teleport out of that fatiguing work and into lower-arousal arenas, which as we have seen are also a source of exhaustion. So it doesn't matter if you're like Dana #1 who finds herself mentally teleporting out of work because she's bored or Dana #2 who is mentally teleporting out of work because she's overloaded—your escape plan to take yourself elsewhere rather than here means that you can't get into a state of flow. All of the digital technologies at our disposal make teleporting when we have too little or too much stimulation easy to do. As MIT professor Sherry Turkle writes about life in our digital world, we have the possibility of being "forever elsewhere."

As we discussed, flow is a state in which we are stimulated, but not too much. We're engrossed in what we are doing and we don't have any need to teleport somewhere else. If we can find flow, we're unlikely to turn to our digital tools to beam us up, out, or to any other place. So the big question is, how do we get there?

Two research teams, one led by professors at the Pennsylvania State University and the other at the University of Maryland, each conducted a series of studies to examine how people might achieve a state of flow when using their digital tools. Their cumulative findings paint a picture of what it takes to be in flow with your technologies:

1. **Flexibility.** You can use the tool in many different ways. For example, if you're a product manager at a software company, you might use a project management tool like Trello to create cards for task management, such as "Access API" or "Create software integration." Or you could not use the cards and instead use color-coded labels to categorize tasks by priority, type, or team.

2. **Modifiability.** You can arrange the tool in a way that will work for you. For example, a teacher might take notes and organize them by modifying their Notion workspace to contain a database for podcast episodes and a content calendar for activities to do with his students in the future.

3. **Experimentation.** You are able to experiment with the features of our tools as you're working. If you're a graphic designer trying different brushes and masks in a tool like Photoshop, you're experimenting with the tool to see how it can aid you in your work.

4. **Playfulness.** You feel like you can be spontaneous, imaginative, creative, and inventive while using the digital tool. For example, if you're an urban planner, you could use a simulation tool like UrbanSim to explore the effects of different city zoning ordinances or traffic patterns on housing choice or urban sprawl.

As these findings make clear, achieving flow when using digital technologies has just as much to do with the characteristics of the tools as it does with our attitudes and approaches toward them. Of course, these characteristics and attitudes must be matched with an appropriate level of work stimulation. If we feel like we can do all four of these things with our digital tools but we're working on an impossible problem, we're not going to be able to get into flow. But the inverse is also true. If we have a problem that falls right at the optimal peak of the Yerkes-Dodson curve but can't orient to the digital technologies we need for the job in the ways described above, we're still not going to make it into flow.

Given these parameters, it's not surprising that of all the studies of flow in digital technology use, the biggest effects are found among video game players. Most modern multiplayer video games provide many different features that allow people to modify their characters and accoutrements. And, of course, most games require experimentation and a spirit of playfulness. Let's not forget that most games are designed with a series of levels that increase in difficulty, allowing players to constantly operate at their skill frontier even as they improve through more playtime.

But it is possible to find flow using digital tools that are much more boring than video games. Recent work has found that people who use social media mindfully have an easier time getting into flow than those who do not, and that when you're in flow on social media you are less likely to experience negative emotions like fear and anxiety. Let's consider a really boring technology, though—a finite-element analysis software called HyperMesh that's used to simulate energy loads in automotive crashes. If reading that description sounded boring, imagine how boring it would be to use it! But even tools like HyperMesh can be used in ways that help us get into flow if approached in the right way. One way that many people in my studies have found flow when using their digital technologies is to create games, impose rules, and develop deadlines. Jensen, a crashworthiness engineer for an auto company, provided one example: "Sometimes if I have a good job, like trying to change the geometry of the front rail (a structural part of a car near the front bumper), I'll make a game out of it. I'll see what are the most creative designs I can come up with [*playfulness*] and I'll kind of mess around with HyperMesh to see if there's new ways of rendering the shapes that I hadn't considered before [*experimentation*]. If I do that, I can get just sort of lost for hours on the design. It's like I just got to work and it's already lunch." Or consider the strategy used by Darcy, one of Jensen's colleagues: "If I've got a bunch of design changes to make, a lot of times I create these little test modules inside HyperMesh that are basically a whole set of commands that will automate things like putting in section cuts [*flexibility*]. Then to keep it interesting I'll rearrange the menus and render the parts in different colors just so I can see the changes I'm making better [*modifiability*]. I love doing those things because it just makes the day fly by." Jensen and Darcy don't teleport out of their work when they create gamelike conditions in HyperMesh. They are present, engrossed, and working at a level of arousal that keeps them on the frontier of their skill level. They are here, not elsewhere.

So where does that leave the two Danas we met earlier? Dana #1 found two ways to get into a state of flow using her digital tools at the PR firm. The first way was by asking her boss for more challenging work. "It took some courage, but I told her I could be handling things that were more complex, and she gave me a shot at it. Not everything is better now, but some of the things she gives me require more skill and thought, and I feel much more energized working on those." The second way was by changing how she oriented to her tools for the tasks that still put her to the left-hand side of the Yerkes-Dodson curve. As she told me, "I've basically started exploring different ways to generate content in the tools we have. There is this one app that we use that you can run quick analysis of story content. I started to play around with the features in there to see if I can run different tests that give me different insights on what an editor might be interested in given their past pubs. I run these little experiments, and I've created these test beds. It's kind of cool. Time flies when I'm doing that." By being courageous with her boss and making some small changes in her digital technology use, Dana #1 was able to find the same kind of flow at work that made Mariana so happy and report such low exhaustion.

Dana #2's story doesn't end so well. She quit. "I just felt like I'd had it," she told me via phone after I learned she had resigned. "I was too burned out from all the complex work and all the new technologies they kept making us learn. If I could have just figured out how to get back to that place where you lose yourself in the work, I think I would have had the energy to stay. But I couldn't get there." Her story provides a cautionary tale. We've got to put in the work to find ways to get into that state of flow so our Level 1 exhaustion stays low. If we're at either end of the Yerkes-Dodson curve for too long, we can transition to Level 2 exhaustion and be unable to recover. Of course, just changing the way she used her digital tools at work likely wouldn't have been enough to stop Dana #2's Level 2 exhaustion. To do so, she would have also needed to figure out how to disconnect better from work so she could more fully recharge. That's what we'll explore next.

FINDING COMPLEMENTARY OPPOSITES

Remember those entries labeled "tiling" on Mariana's calendar? They teach a second important lesson about being here, not elsewhere: Finding flow in activities that don't require the use of digital technologies can reduce the overall amount of time we spend in front of a screen, thus giving us more energy for when we do need to sit down and work with our digital tools later.

As I discussed in the introduction, there is mounting evidence that digital detoxes and technology abstinence do not produce significant increases in people's well-being. The goal is not to move away from our tools permanently or for extended periods, but to take advantage of any periods of respite that we do have from them so we can recharge and feel energized when we return to our screens.

Researchers of digital media use who focus specifically on feelings of exhaustion have found that voluntary, purposeful short breaks from digital tools can temporarily reduce one's feeling of exhaustion. For example, a study of working age adults in Korea found that when people spent their one-hour lunch break taking a walk outside away from their phones they reported significantly "increased vigor and reduced emotional exhaustion" compared to people who spent their one-hour work break on their phones doing non-work-related activities. When we look at the data specifically about reductions of exhaustion associated with voluntary breaks from our digital tools, they tell a consistent story: We are best able to recharge when we have some time away from our devices.

As you might imagine, there is no shortage of studies that try to figure out what the best non-technological activities are for helping us to recharge. There are many suggestions. But here are some of my favorites:

- **Spending time outdoors.** Research shows that spending time outside without digital tools at the end of a workday increases positive emotions and reduces feelings of exhaustion at the start of the next workday. But—and this is important—it only works for people who reported truly "connecting" with nature when they were outside.

- **Looking at the water.** Spending 1 minute and 40 seconds looking at a body of water significantly reduced systolic blood pressure relative to diastolic pressure by a significant amount when compared to looking at a tree or the ground. Reductions in heart rate were also greater when people looked at the water as compared to the ground, as were subjective assessments of relaxation and rejuvenation. The wider the body of water that someone looked at and the more intensely they looked at it, the bigger the effects.

- **Having sex.** In a result that will surprise probably no one, married people who lived together and had sex in the evening after work reported significantly higher positive emotions at work the next morning when compared to people who did not have sex. They experienced a 5 percent increase in positive mood each time they had sex the night before. Most importantly, the results also showed that the more participants rated that they felt engaged in the sex, the stronger its effect was on positive mood.

All these studies find that people recharge best when they *engage* in activities that do not involve their digital technologies. And the more engrossed they are in those non-digital activities—whether that's taking in nature, contemplating water, or having sex—the better they recharge.

OK, so let's get back to Mariana's tiling. The reason this activity was so effective in helping her to recharge after experiencing daily Level 1 exhaustion was that it was engrossing for her—it enabled her to enter into a state of flow doing something that didn't involve using technology. In fact, the more into flow Mariana feels while tiling, the less she wants to use her digital tools. As she described to me: "This kind of weird thing happens to me when I'm tiling or doing whatever hands-on project like that. I just don't want to check my email, or Slack, or the news, or anything digital really. Like, it turns me off. Normally I can't stop picking up my phone. But, honestly, the more I'm into a project, the less I care. And when I'm done tiling for the day, I don't even want to go on Twitter or watch a YouTube video then. It's like I'm over it."

I've found that people like Mariana who are able to get into flow in non-digital areas of their life don't randomly look for non-digital activities. They find activities that I call "complementary opposites." A complementary opposite is an activity that is roughly the opposite of your day job in terms of the physical skill you need to do it, the medium in which it's done, the location in which you carry it out, and, of course, the technologies you use when doing it. But it's complementary to what you do in your job when it comes to the kind of analytic reasoning and critical thinking needed. I picked Mariana's tiling to illustrate in this chapter because it's a complementary opposite that she and I share, so it's easy for me to describe.

The kind of work that puts me in a state of flow and makes my digital tools seem unattractive to me is home renovation. In my day job as a professor, I am at my computer for nearly eight hours. I receive hundreds of emails in a day, and I'm searching for research articles on Google Scholar and Web of Science and reading them online and in PDF format. I'm using statistical analysis tools like R and qualitative analysis tools like ATLAS.ti. I'm on Zoom with students and leaders of companies. I'm updating my courses' Canvas sites. I'm creating presentations for various audiences in PowerPoint and Keynote, writing papers in MS Word and Google Docs, and using AI tools like Chat-GPT, Claude, and Gemini for my own use and to learn more about them for my students and clients. But when I'm hanging drywall, doing electrical work, installing shoe molding, or like Mariana, tiling, there are no digital technologies in sight. There are also no students, or clients (other than my wife), or peer reviews, or papers to read about "transaction costs" or "structuration." Doing home construction projects is about the furthest from the academic life you can get. They're practically opposites.

But they're also complementary. To do a research project that culminates in a peer-reviewed article or to develop a course that is really meaningful for students requires a lot of planning, problem-solving, precision, and revision. It's mostly a solo effort, but you always need help from people along the way. You don't really have a "boss," and you stop collecting data, revising a paper, or preparing for a lecture when *you* think it's good enough. But people will interact with your final product and judge how well it meets their needs. All of those characteristics also describe home construction projects. There is a complementary set of cognitive skills in both lines of work. If I had to start totally from scratch with the physical skills (e.g., learning how to feather a drywall seam) and cognitive skills (e.g., planning how to hang the drywall), it would likely be too much for me and I wouldn't be able to get into flow. As Mariana told me, "One thing I love about tiling is that I can still use the professor part of my brain, but I don't have to do professor work." I totally get that: complementary opposites.

I've heard of many other examples of complementary opposites over the years. Of course, what is "opposite" and what is "complementary" will vary from person to person. But as a few examples:

- Accountant & cooking

- Event coordinator & rock climbing

- Hedge fund investor & jiu-jitsu

- Structural engineer & weight lifting

- Architect & car restoration

- Lawyer & salsa dancing

Like Mariana and me, when people find their complementary opposites, they report higher levels of flow, find they are less interested in going back to their digital tools, and report lower levels of exhaustion (something I actually do measure, as you know).

Disconnecting doesn't lead directly to rejuvenation. As Shawn Achor, author of *The Happiness Advantage*, observes, "Most people assume that if you stop doing a task like answering emails or writing a paper, that your brain will naturally recover, such that when you start again later in the day or the next morning, you'll have your energy back." But as Achor shows, it doesn't if you're mentally fatigued or suffer from digital exhaustion rather than physical exhaustion. Cognitive absorption in another task in which you are engrossed is a significantly better way to recharge than pursuing leisure activities where your mind is disengaged. Laura Giurge and Vanessa Bohns, whose research we read about in Rule #4, found that making plans to do non-work activities actually helps people to recharge better than when they do nothing or stumble into activities that are not complementary opposites. In one study, they found that people who set goals for their days off from work were 12 percent happier than those who didn't make plans. In another study, they found that people who made plans for their weekends were 13 percent happier than those who did not. And in a third study, they found that people who made plans to engage in specific activities in the evening after work were 10 percent happier than those who did not. As they conclude, "Passive 'rest and relaxation' isn't as effective for recovering from the daily grind as using breaks to accomplish your goals—not your work goals, but your personal goals."

Another study drawing on surveys conducted with working adults over a seven-month period found that people who reported the lowest exhaustion and highest energy levels after non-work activities were those who engaged in an activity that was very different from their work activity and who took the activity very seriously. If the activity was too similar to what they did at work or they did not take the activity seriously, they reported higher levels of exhaustion and lower levels of energy when they returned to work.

Another issue is worth exploring: How thoroughly should you plan your non-work activities to recharge? The findings of thirteen separate studies showed that people who schedule their non-work leisure activities too precisely by planning to begin and end them at a specific time report less enjoyment and less engagement while they're doing them. The studies suggest a better strategy is to "roughly schedule," meaning without prespecified times—like, "I'll work on the backsplash tile in the morning," as opposed to deciding "I'll start right at 9 a.m." Putting this all together, if you're looking to find flow in your non-work activities and to be in the moment, you should:

- Identify an activity that is a complementary opposite from your work.

- Make sure it is something in which you can engross yourself.

- Choose something that is cognitively engaging rather than pure rest—and it should not involve substantial use of digital technology.

- Plan when you're going to do it.

- Don't schedule it precisely but decide on a rough time frame to begin and end.

Engaging in serious activities that are opposite from what we do at work but that complement that work in important ways can bring benefits of rest, relaxation, and renewed energy. Importantly, these activities also keep us willingly away from our digital technologies, at least for a while, in ways that feel comfortable and natural. As author Anne Lamott wisely observes, "Almost everything will work again if you unplug it for a few minutes, including you."

PART III

::::::::::::::::::::

Complex Contexts

CHAPTER 4

How Not to Be an Energy Vampire: Lessons for Workplace Managers

G riffin is a partner at a boutique architecture firm. He's a workaholic, and he knows it. But he's also sensitive to the fact that not everyone in his firm cares to work as many hours as he does. About a year before I met him, Griffin started including a little message below the signature line in his email and in his profile on Slack. It read: *I'm sending this message now because it's a good time for me. Don't feel you need to respond to this message right when I send it if it doesn't work with your schedule.* Yet despite his reminder that recipients of his messages need not respond immediately, Griffin found that employees at his firm nearly always responded right away, even late at night and on the weekends. "I don't get it," he lamented to me. "I've been trying to make a change to make our culture better, but it just doesn't seem to be working."

In the last couple of years, I've seen a big shift in the awareness of digital exhaustion by workplace managers and senior leaders. The chief product officer at a SaaS (software as a service) company recently told me, "Ten years ago, or maybe even five years ago, I wasn't sure digital exhaustion was really a thing. But now it's obvious that people are feeling it in a big way. We've got to do something about it as a company." And a mid-level manager at a biotech firm noted that since the pandemic, "I've really noticed that people are just more and more overloaded by their technologies. We've got to be more sensitive about this in our org and figure out ways to help them."

Unfortunately, many of the well-intentioned fixes like Griffin's note to his employees just don't work. As Harvard professor Dorothy Leonard showed many years ago, most employees don't give a lot of credence to what their leaders *say* about technology-use norms in an organization. Instead, they pay attention to what their leaders *do*. Because Griffin was always quick to respond to other people's messages, he set the tone for the rest of the company that other people should too. Rana, a project captain at Griffin's firm, noticed this discrepancy between what Griffin said and what he did: "Sure, he says we don't need to respond right away. But he's always responding to us right away, so it's clear what the priority is. It just makes his little jingle—or whatever you call it—feel disingenuous." Rana's not alone in feeling a disconnect between what leaders say and what they do when it comes to dealing with digital exhaustion at their companies. Consider the following commentary by employees at other companies who saw notes from their own bosses that were similar to Griffin's:

"If I have to see one more person tell me not to respond to an email just because they sent it, I'm going to scream."

"It's nice that management is aware that we're all getting burned out, but those words aren't going to fix it."

"That shit is such bullshit. I mean, you can't just make everyone feel less overloaded by telling us not to feel so overloaded. It's like, you're the boss—of course we're going to follow your lead."

Cal Newport suggests that managerial approaches to digital exhaustion like the ones that precipitated these responses are really just attempts to skirt the issue, and that rather than fix the problem, they exacerbate it. He writes, "It's this mindset that leads to 'solutions' like improving expectations around email response times or writing better subject lines. It leads us to embrace text autocomplete in Gmail, so we write messages faster, or the search feature in Slack, so we can more quickly find what we're looking for amid the scrum of back-and-forth chatter. These are the knowledge work equivalents of speeding up the craft method of car manufacturing by giving the workers faster shoes. It's a small victory won in the wrong war."

The latest managerial "solution" that is just coming into vogue, and that I fear will be here for some time to come, is to herald the coming wave of generative AI tools as the definitive solution to the problem of digital exhaustion. Many companies that develop GenAI tools or applications have begun to aggressively advertise their products as the fix for digital exhaustion. Microsoft's Copilot (which is powered by OpenAI, the makers of ChatGPT), as one example, does boast some impressive stats when it comes to helping employees manage the onslaught of communications coming at them every day—at least according to its own analyses of its employees and select customers. As Microsoft's VP for its Modern Work initiative, Jared Spataro, says, "Over the past year, I've personally experienced the significant impact of #Microsoft365Copilot on my own email. From summarizing long threads to drafting responses and even answering the question 'what's hot in my inbox right now?' Copilot collaborates with me to save time and effort. I don't want to go back to life without it." That sounds pretty impressive. We'll talk about AI specifically in chapter 6 and explore how using it in certain ways may be helpful for managing our digital exhaustion. But as we learned in part I of this book, we can't deal with our attention fragmentation, inference-making, and emotional responses to digital technologies simply by using AI. In fact, as we'll discuss in the last section of this chapter, if we're not careful about how we deploy AI in our organizations, we could really make digital exhaustion much worse.

So, as a manager and leader in your organization, the key to reducing digital exhaustion for your employees is not to simply encourage people to develop their own healthy habits or to implement even more digital tools to solve the problem. Instead, it is to create a healthy culture of technology use within your organization and to make sure that you are not an energy vampire who uses digital tools in ways that exhaust others. Let's explore how to adapt the rules we learned in part II to help others manage their digital exhaustion.

ONLY YOU CAN PREVENT TECHNOLOGY PROLIFERATION

When I met Gunter, he was three months into a job as the chief technology officer of a midsize company that developed infrared imaging devices. As part of his "listening tour," as he called it, he talked to people in different divisions to understand what their technology needs were and how they thought about the company's current portfolio of digital tools. What he learned shocked him: "People in nearly every department would show me all these tools they were using. There were so many—many more than we had on our official vendor list. I'd ask them why they were using some particular technology, and they'd say, 'Oh, because it helps us.' Then I'd ask them how they got it, and they'd say, 'I don't know, I think my manager just got it for us.'" Gunter decided to do a formal audit to identify how many different digital tools were actually being used across the company. It turned out it was more than 150. To make matters worse, tools like Slack were being paid for multiple times and one group's Slack accounts were not connected to another's.

There were two interrelated reasons why there were so many digital tools being used at the company. The first reason was ideological. Many managers believed that the solution to people's communication, information, and digital exhaustion problems was that they didn't have the *right technologies* in place. As one of the team leads at Gunter's company confessed to me, "I just thought we needed to communicate faster and that if we could get more tools that would allow us to be more 'async,' we'd be more efficient. That's why I got Slack." I've heard this rationale from managers more times than I can count, yet it's what so many leaders I've talked to seem to believe—that the solution to our digital exhaustion is to use more digital tools. Perhaps it's because that's what the digital tool companies sell us: a promise that if we use their "perfect" tool, we'll be freed from the other tools that just don't work. But this way of thinking directly violates the principles we learned about in Rule #1 (Stop Using Half Your Tools), Rule #2 (Make a Match), and Rule #4 (Wait). It relies on the assumptions that more tools are the answer, that one type of tool is good for every job, and that faster is better.

The second reason was practical. This manager could simply subscribe to Slack for his team because Slack's pricing and sales model allowed him to do so without having to get formal approval. Many digital tool makers have figured out a loophole that allows them to get around the lengthy process involved in enterprise technology sales. Most managers have authorization to spend a certain amount on their corporate credit cards without incurring scrutiny. Many digital technology companies, especially those that make subscription-based software, exploit this fact by pricing their software to fall below the threshold that would require review and prior approval by most companies. Also, many SaaS companies have developed an explicit "land and expand" sales strategy in which they first look to sell to individual managers within a company. Then, once they have enough teams using their tool, each with individual subscriptions, they approach the company through its official technology purchasing channels and basically say, "You already have one hundred teams paying for our software, why don't we create an enterprise account?" What makes for a subtle and effective sales strategy for digital technology providers often results in technology proliferation.

The problem, as we learned in Rule #1, is that it's often hard for any one employee to stop using a tool that is widely adopted at their company. Your employees can't easily opt out of MS Teams if it's their team's primary way of communicating. That means it needs to be a manager or a senior leader's job to stop digital technology proliferation at the company.

Gunter understood his responsibility and acted boldly. He instructed the company's accounting team to identify all recurring fees charged to technology vendors on corporate cards and to flag any new charges for review. Gunter then canceled each subscription that was not authorized by IT. If the manager wanted to reinstate it, he or she had to make a formal request in writing. This was a hardline approach for sure. But it really forced managers to question whether they really needed the technology or not and to confirm that it was really the right tool for the job. Six months after the big cancellation, the company was using thirty fewer tools. "I took some heat for that," Gunter told me. "But in the end, people routinely tell me they're happier because they either didn't realize what a burden it was to keep up with so many tools or that they wanted to stop but didn't know how." Sometimes only senior leaders have the authority to create enough friction to stop the bad choices from being the easiest ones.

It's also important for leaders to model the type of technology-use behavior that reduces digital exhaustion. Telling employees not to respond to emails after hours if they don't want to isn't enough. You can do this by setting clear guidelines for what types of digital tools are appropriate to use for what kinds of work. Then, you can model this behavior by following those guidelines when you interact with your team. Several leaders that I've worked with who have excelled at creating a healthy culture of technology have developed cheat sheets for employees to keep by their computers to remind them of what kinds of tools to use for what purposes. They even encourage their employees, if they are feeling insecure, to simply say to someone: "The boss's cheat sheet says this is a telephone-worthy conversation." Other leaders have also done the same thing for Rule #3 (Batch and Stream), Rule #4 (Wait) and Rule #8 (Be Here, Not Elsewhere), providing guidelines on appropriate lengths of time to allow for responses to coworkers and customers and outlining what tasks should take priority over a response. It may seem strange at first to provide such guidelines since most of us like to take pride in being good communicators. But it's foolish to think that what constitutes "good" is the same in every context. Making clear that there are well-articulated and visible norms on your team is a surefire way to help prevent digital exhaustion.

STOP TALKING SO MUCH ABOUT TECHNOLOGY

"I know it sounds insensitive and trite," Misha began in response to my question about changes going on at her company, "but if I could wish for one big change in the world it would be that everyone just stop talking so damn much about technology. It's too much!" I understood why she felt that way. I'd spent the last two weeks at her company, a commercial airline services firm, and I too was overwhelmed by the number of discussions I heard about digital technologies. Almost every team meeting I participated in included some talk of a new digital tool that the team was using. Senior leaders seemed to talk almost nonstop about hybrid work and whether they had the right technologies and strategies in place for it. I sat in on four meetings in which division directors talked to their managers about AI and its role in helping to streamline operations at the company (though no one was using AI yet, as far as I could tell). And the company's internal social networking tool was filled with industry articles about the technological changes that were shaking up the airline industry.

As we discussed in chapter 3, too much talk about technology can be exhausting. It is particularly exhausting if those conversations about technology follow a deterministic narrative and make people feel as though new technologies will change everything and that they're just along for the ride without any say in where they will go. Of course, many of them do follow such a narrative. Moreover, as we learned in Rule #6 (Act with Intention), feeling powerless around your technologies can lead employees to act without intention and get sucked into the trap of feeling like they haven't accomplished much after long bouts of use, which then further enhances feelings of exhaustion. After spending twenty years in and out of hundreds of companies, I can say this with great certainty: Your employees think you talk about technology much more than you think you do. And it's exhausting them.

In addition to the sheer volume of managerial communication about new technologies that most employees experience, the things senior leaders say about digital tools linger in the organization for quite some time and can have nefarious and long-lasting effects. One of my early research projects was with General Motors. At the time, the company was working hard to digitally transform product design by shifting analysis of vehicle performance to a host of technologies that would allow engineers to home in on design solutions more quickly. As part of that process, they were implementing a new digital tool they called CrashLab that would automate the way engineers set up their simulations of crash tests. It just so happened, for a variety of reasons that are not all that interesting, that one of the groups of engineers I was studying heard a lot of talk from their managers that CrashLab would *speed up* their analysis. A second group of engineers doing the exact same work as those in the first group didn't hear that CrashLab would speed up their work. Instead, they heard that it would *standardize* how they built their models. Because I followed these two teams closely, I was able to actually compare how many "speed" messages or "standardization" messages each group heard. The first group heard six times the number of "speed" messages as they did "standardization" messages. The second group heard four-and-a-half times more "standardization" messages than they heard "speed" messages. Remember, both groups were doing the exact same kind of work and were about to use the exact same digital tool. But the things they heard about CrashLab from their managers were quite different. Here are two examples:

WHAT GROUP 1 HEARD: CRASHLAB = SPEED	WHAT GROUP 2 HEARD: CRASHLAB = STANDARDIZATION
Manager: Apparently, CrashLab is supposed to help make sure we stay ahead of the development curve. It sounds like the barrier positioning and accelerometer placement functions work really good and will save you lots of time, so make sure to make use of them. **Engineer:** Do you know how the algorithm works behind that, I'm just curious? **Manager:** Behind what? **Engineer:** For the automation. **Manager:** I don't know, but do you remember Brett Pascal who used to work in [another vehicle program group]? I think he was involved in it somehow so you could ask him. Anyway, you should be able to figure it out with the training and then we should see you work like lightning.	**Manager:** So, I know what you're thinking about using CrashLab. Why should you do it when other things work fine and it's only going to slow you down? Yeah, CrashLab might be a bear in the short run, but a short productivity loss is nothing compared to better benefits in the future. **Engineer 1:** That's pretty dramatic there! **Manager:** You're right, but I'm just saying. **Engineer 2:** So we really need to use this? **Manager:** Just scale up to it. Figure out how to use it and migrate your work over. Then it won't be such an impediment on it. It's really going to help you guys standardize so you can compare your results better.

So what happened? I tracked the use of CrashLab by each engineer in both groups weekly for fifty-one weeks. The engineers in both groups did exactly what you would expect good employees to do. They tried CrashLab, and they looked for the improvements that their managers told them to expect. Engineers in the first group ran a bunch of tests to see whether using CrashLab sped up their work more than simply doing their work using the usual methods. Engineers in the second group didn't run speed comparisons. Instead, they looked to see if their coworkers were using CrashLab's features in the same way they were so that they could set up their models similarly and consequently be able to compare their outputs easily with one another.

Engineers in the first group found that CrashLab didn't actually help them run their analyses any faster than they could without it, so they largely stopped using it. But that decision caused them a fair amount of stress and exhaustion. As Clara, an engineer in the first group, told me, "I really wanted to try to find a way to save CrashLab. Stan and Dennis [her two bosses] really wanted us to use it. But I just couldn't justify it. It was just slower. And the whole reason we're supposed to use it is to speed things up." Then she continued, "It's kind of stressful that it's not working out like they wanted, and I'm feeling kind of anxious about it. But I guess that's that." I could understand Clara's point. She felt that she was letting her bosses down by not using CrashLab, and she felt slightly insubordinate because her speed comparison tests had basically proven them wrong. Many of the other engineers in this first group who heard all of management's talk about "speed" felt the same way. And not surprisingly, they scored high on the digital exhaustion scale. Engineers in the second group had a much easier time. They too did comparisons—but not to determine if Crash-Lab helped them set up their models faster. No one told them that it should, so they didn't have that as a goal. Instead, they paid attention to whether CrashLab helped them set up their models similarly to others in ways that aided engineers in making comparisons. It turns out CrashLab was pretty useful for that. Engineers in the second group didn't find themselves in the uncomfortable position of proving their managers wrong, and most submitted low exhaustion scores.

There is a considerably robust line of research on the power of managerial framing around new technologies. It all points in the same direction: Employees approach new technologies in the workplace and evaluate their experiences using them based on what they've heard managers say about them. Unfortunately, that line of research shows that managers most often don't know what they're talking about. That's not a knock on them. The impact of new technologies is just really difficult to be able to predict. As my amazing colleague Steve Barley, one of the foremost authorities in the world on technology and organizational change, writes, "After nearly forty years of studying work, technology, and organizing, I have concluded there is only one certainty about technological change: You almost never get *only* what you expect and sometimes you do not even get that. However, something usually happens." As it was with Steve's, my studies have also shown that managers don't often get it right because technological change is such a messy process—too messy to be able to predict its outcomes with good accuracy. That's not their fault. But it does suggest that managers might benefit from cooling it a bit on their predictions about technology's yet-to-be-seen effects, or at least dialing down their level of certainty when discussing them with employees. As we've seen, if what you say about technology clashes with your employees' own experiences with it, you can create more exhaustion than relief. And whatever you think a technology's effects will be, you're more likely to be wrong. So save everyone the trouble and just talk about technology less.

RETHINK HYBRID WORK:
COORDINATION AND CAMARADERIE

After email, hybrid work and AI are the two technological changes to work over the last half century that have created the most opportunities for employees to feel digitally exhausted. I won't bore you with the stats on the magnitude of the global shift to hybrid work. Those shifts are now well-documented even if policies around hybrid work remain in flux for many companies. Although hybrid work can connote several different things, I refer to a hybrid workforce as one in which employees are working from many different locations at any given time. Some may be working in a building at the company's headquarters, while others are working from home, a hotel, the beach, a client site, a satellite office, or countless other possible locations. Just about all work today is powered by some form by digital technologies. But it's the combination of digital devices, software applications, and networked infrastructure that make it possible for people to work together at a task level at considerable physical and temporal distances from one another.

Since the pandemic rapidly accelerated the great work-from-home experiment, many companies have realized that people can be as productive working from remote locations as they can be in the office. But when we make the mistake of treating all types of work as equal and assuming people can work remotely all the time, we increase the likelihood that they will experience digital exhaustion. Likewise, when we try to bring people into the office when they don't need to be there, we fail to take advantage of the capabilities of digital technologies that allow people to have more flexible work schedules and more focus. The randomness associated with these decrees about when to be in the office and the mismatch of work expectations with digital technologies is a major source of exhaustion for people working remote and hybrid arrangements.

Today, you can hardly flip through a newspaper or scroll through your LinkedIn feed without seeing a headline asking some variation of: "How many days a week in the office is ideal for hybrid work?" As if there were a magic number that could fit all situations. This is the wrong question. The better question to ask would be: "When should employees come into the office?" When digital exhaustion is a top criterion for consideration, the answer is quite clear: Have your employees come to the office when the demands for coordination are high and/or they need to build camaraderie. Let's look briefly at each of these.

Coordination

As a manager, your job is to help employees make the right match with their technology and communication needs. Don, who runs operations teams at an online e-commerce site, manages more than twenty teams whose members are geographically distributed. He works with the team leads to map out projects. When those projects include tasks that will require high degrees of coordination, he tells team members to come into the office. When the team shifts to tasks that require lower degrees of coordination, he lets them choose if they'd like to work from the office or elsewhere. The key to Don's management success lies in mapping project phases to determine task coordination needs and planning far enough in advance to give employees time to get to the office or choose their work location.

When I work with companies to help them plan their remote work in ways that reduce digital exhaustion, I use a framework that comes right from Rule #2 (Make a Match). If you're working with someone in the office synchronously, you're working in the same place and at the same time as your colleagues. But if you're working asynchronously and out of the office, you're working in a different place and at a different time from your colleagues. From my own research, and drawing on the sizable research on remote work coordination by other scholars, I've seen that the most important factor for reducing digital exhaustion in remote and hybrid work is to think about the lessons we learned in Rule #2 carefully when we are trying to decide if our workers need to be together in the office or not. I find that it's helpful to think about richness at a higher level in this context, focusing simply on whether the work that we're doing can be done asynchronously or synchronously. If the tasks that we're working on are sequentially interdependent, then our coordination needs are low. That means we can do our work asynchronously, which is typically easy enough to accomplish if the team is remote. In fact, getting a team together whose tasks are sequentially interdependent—whether through digital tools that afford synchronous communication or in person in the office—might be too much of a distraction. If the work that we're doing is characterized by pooled interdependence, we probably need some ability to communicate synchronously to resolve issues where the parts of our tasks come together, but we're likely going to be fine if the team is working remotely. However, if we are doing tasks characterized by reciprocal interdependence in which we need a constant back-and-forth to work out ambiguities and resolve uncertainties, we want to have synchronous communication, and it sure helps if we are together in the office. For these kinds of tasks, nothing beats seeing our colleagues, reading the context of their concerns, adjusting to their pace and emotions, and really getting on one another's wavelength. All of that happens best when we're in a room together.

To add some numbers to this, my research team and I conducted a large study of nearly 150 projects across eighteen companies. We tracked the complexity of the teams' activities. We also collected log data about what technologies employees used while they were working on each activity and from what locations they were working. Then we surveyed managers to learn how well the teams performed, and we surveyed team members to learn how engaged they were during each activity. We found that when managers helped teams to match the task's coordination needs with the appropriate technologies (both digital and analog, like the office), both the team's performance and their engagement were significantly higher than if managers failed to make the right match.

Camaraderie

Coordination need isn't the only thing that Don is looking for when he maps out his teams' workload to decide where they should work. A second critical variable is camaraderie. Here, I use "camaraderie" in the broadest sense to include feelings of goodwill, friendship, trust, and a shared set of norms and expectations. Camaraderie connotes a feeling of oneness or union with a team or organization. Nearly three decades of research on distributed work shows that teams that work virtually or remotely struggle to develop these aspects of camaraderie. It's not impossible to do by any stretch of the imagination; it's just harder to do via digital tools than it is in person. Pamela Hinds from Stanford University and Catherine Cramton from George Mason University showed that even teams that do a relatively good job developing camaraderie through smart uses of their digital tools end up getting a major boost in feelings of trust, respect, and general affinity for their colleagues when they meet face-to-face in the office periodically. My frequent coauthor Tsedal Neeley, who is an expert on global work, writes in her book *Remote Work Revolution* that teams should hold in-person launch sessions to help people develop camaraderie, and then relaunch several times throughout a project.

Employees who rarely see one another in person report lower levels of trust, affinity, and understanding than those who see one another frequently in the office. As Don told me, "My teams do better in terms of morale when they connect in person. They just like each other more. So I make sure to plan random get-togethers for them so they can bond." As one example, Don took two of his teams, comprising sixteen people total, to San Diego for two nights. They discussed some aspects of their project during the off-site, but most of the time was spent at restaurants and playing golf. As Don explained, "The goal is to just get people together, and it's good to do it in a neutral location where everyone has to travel to get there. I've had to rethink my budget asks to finance these trips for the team, but they certainly pay off in big ways. The goodwill carries after, and my teams get along better because they know each other better."

Tsedal and I have found in our joint research that there are indeed ways that good managers can boost feelings of camaraderie by encouraging employees to use digital tools if bringing people together in person is not a viable option. We examined two companies where employees were using internal social tools for knowledge sharing. Senior leaders encouraged employees from different divisions to regularly share interesting facts and updates from their personal lives on the company's social media platform, similar to how they would on Facebook or Instagram. The idea was that by discovering coworkers with shared interests or similar backgrounds outside of work, employees might feel more comfortable reaching out to those individuals about work-related matters. Initially, employees felt awkward posting what they dubbed "Facebook-like" content on a workplace tool, but management not only encouraged this behavior but also modeled it themselves. Soon, the platform's algorithm began connecting people with similar non-work interests. The company found that when employees engaged in conversations about food, sports teams, movies, and fitness, they were more likely to discover new things that they did in their work that would be useful for them, and they began to reach out to one another for work-related questions as well. One employee who had bonded with a coworker over a shared love of independent films noted, "Talking about movies with her made me feel comfortable asking for advice on some tricky work matters."

Often, employees need a little conversational nudge to approach a distant colleague and turn them into an occasional collaborator. Both work-related and non-work-related content on digital tools can act as that social lubricant, facilitating these conversations. Importantly, employees can maintain lightweight relationships with their sporadic collaborators by following and commenting on their posts during times when they don't need help. This ongoing interaction makes reaching out for assistance feel less transactional and more organic. In large organizations, this also helps employees feel connected to their company and view themselves as part of the community. As a manager, you have to encourage and facilitate these kinds of informal, personal connections because, as Tsedal and I discovered, without active encouragement by management, employees often stop using their tools in this way.

GET INTELLIGENT ABOUT ARTIFICIAL INTELLIGENCE

As we previously discussed, it's rare for managers to make good predictions about how new technologies will change work. Trying to do so is likely to be even more of a fool's errand when it comes to generative AI tools that are designed to learn and change their own capabilities with some regularity. That makes it even more important for us to consider how to approach AI in the workplace so that the uncertainty these tools bring, mixed with the apprehension many employees already have about them, doesn't lead to widespread exhaustion. Although we'll talk specifically about AI in chapter 6, I think it's worth spending some time here to discuss how managers and organizational leaders can think about implementing AI in ways that don't burn their employees out.

Because I was seeing employees' exhaustion rates climbing dramatically as companies scrambled to figure out how to incorporate AI into existing workflows, I decided to start a project with ten knowledge-intensive companies using AI to try to find ways to mitigate the exhaustion employees felt when grappling with these new kinds of tools.

The STEP Framework for Managing the Introduction of AI

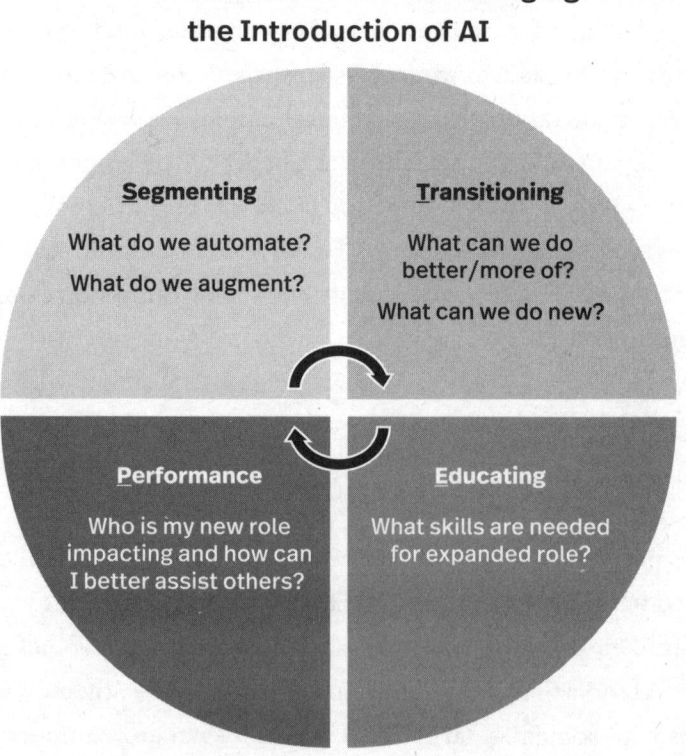

Segmenting
What do we automate?
What do we augment?

Transitioning
What can we do better/more of?
What can we do new?

Performance
Who is my new role impacting and how can I better assist others?

Educating
What skills are needed for expanded role?

In this project I developed a framework for leaders to help employees harness AI's capabilities to improve their work and benefit their organizations without fomenting feelings of exhaustion. This framework, called STEP, consists of four interconnected activities: **S**egment AI's role for task automation or augmentation, **T**ransition activities across roles, **E**ducate workers on using AI's evolving capabilities, and evaluating **P**erformance in ways that reflect their new learning and the new levels of support they provided to their colleagues as their work roles changed.

The idea of the STEP framework is to actively involve employees in figuring out how to use AI in their work and engage them in helping to decide how AI should change the way we organize. We want employees to be present and engaged when using AI and to use it with purpose. I've found that the STEP framework helps them to do both.

To illustrate how you can use STEP effectively to roll out AI in your organization without pushing employees deeper into the pit of exhaustion, we'll look at examples from three companies who have adopted it successfully (I'll use pseudonyms for each one): a medical device manufacturer, "HealthCo"; a marketing agency, "MarkCo"; and a metropolitan planning agency, "UrbanGov." Specific departments in these organizations used the framework to leverage their employees' expertise with AI's capabilities. These lessons provide examples about how to implement the framework to improve employee experiences and forestall exhaustion.

Segmenting Tasks

There's a lot of talk about AI taking over jobs, but this overlooks that most jobs in knowledge-intensive companies involve multiple tasks. No single AI will perform all the tasks of one person's role. Leaders should ask, "How will AI affect the various tasks my employees conduct?" Estimates suggest up to 80 percent of the US workforce could see at least 10 percent of their tasks affected by AI, with 19 percent potentially having 50 percent of their tasks impacted. AI won't touch all tasks, but it can and likely will significantly influence many. Leaders should enable workers to segment tasks into three categories: 1) tasks AI can't or shouldn't do, 2) tasks AI can augment, and 3) tasks AI can automate.

Take HealthCo's ethics and compliance division, which adopted ChatGPT for its junior staff. Leaders at HealthCo and other companies encouraged workers to lead the segmentation process, fostering trust and showing that automation wouldn't eliminate their jobs. Initially, leaders in the ethics and compliance division encouraged staff to determine which tasks AI would not be helpful for. Tasks such as determining compliance with federal policy and deciding parameters for safeguarding the company's IP when working with outside consultants quickly rose to the top of the list. These tasks required human judgment and contextual understanding that AI lacked.

Next, the team identified tasks where AI could augment their work. One time-consuming task was ensuring that contracts accurately reflected RFP details. Here, AI was very helpful. By reading through an RFP and a standard contract template, AI could generate a draft contract that reflected the terms of the agreement. Paralegals then reviewed and revised these drafts, focusing on specific areas of concern that they could identify thanks to their experience.

Finally, it was time to identify tasks that the AI could automate completely. One such task prime for AI-only execution was the laborious job of writing emails to outside parties requesting contract changes. After segmenting tasks into these three buckets, employees got to work figuring out how to use the AI to augment and automate the tasks they had identified.

Leaders at HealthCo and the other companies that used AI effectively encouraged workers to take the lead on the process of segmentation because they understood that these workers would be in the best position to evaluate where and how AI could help. They did so by asking employees to experiment with AI tools and by convening meetings in which employees discussed the results of their experimentation and reached consensus on best practices. Employees were happy to be involved in this process for two reasons. First, the fact that leaders asked them to lead the segmentation exercise demonstrated the trust leaders placed in them. But second, and perhaps more importantly, employees understood that automating part of their jobs would not put them out of work.

Transitioning Roles

If AI makes tasks faster and more accurate, employees may have less to do in their current roles. One option, then, is to simply take the work that two people used to do and have one person do it instead, thus reducing head count by one. This is the outcome that people critical of AI worry about. Yet across the ten companies I worked with, only one eliminated jobs or planned to eliminate jobs as a result of efficiencies gained by automating and augmenting work. And the numbers of eliminations were small. Two other strategies were much more common. They involved transitioning work roles by deepening or upgrading them.

MarkCo deepened roles by adopting a chat-based AI to help low-level marketing associates create collateral, like product descriptions and brochures. Freed from routine tasks, employees focused on competitor analysis and campaign testing. Both were tasks the firm did already, but not often and not with the desired level of sophistication.

With employees now having time to transition into these tasks, the firm could build capabilities in these areas in ways their competitors could not.

Leaders began by identifying capabilities that were lacking like competitor analysis and experimentation. They identified employees with the aptitude and interest to deepen their knowledge in these areas and drew on internal expertise to provide initial training. As one of MarkCo's senior account executives commented, "My team hasn't been able to really understand the competition for our clients. I'm thrilled that I can dedicate a certain portion of my team's time to focusing on this area, which is high value, rather than have them spend their time tinkering with formatting on our collateral."

A second way of transitioning work roles is to upgrade them. Upgrading involves having employees perform tasks typically conducted by their managers. For example, at UrbanGov, junior planners traditionally built land-use models. By segmenting tasks, they augmented and automated some tasks, freeing them to develop new modeling scenarios, a task typically done by senior planners. To balance this, senior planners took over relationship management tasks from the lead planner. The lead planner explained, "If we didn't figure out what new tasks senior planners could do, they wouldn't let junior planners get more involved with scenario building . . . After careful evaluation, I decided to give them a big chunk of my job. That allowed them to feel comfortable giving up scenario building. The good news is that now I can focus my efforts in new directions too since I am freed from maintaining all of those relationships."

Across all the organizations I worked with, the key to transitioning work roles was creativity. No matter whether leaders worked with employees to deepen their expertise in certain areas or upgraded their work roles to include tasks that someone higher up did previously, leaders had to envision new tasks that could be done that the company was not doing—tasks that would add value either at the lowest levels of the organization or at the level of middle management.

Educating Workers

Both segmentation and transition demand that employees learn new skills. Some of those skills are directly related to using data, algorithms, and AI. Employees need to know how new AI tools work, how to train them on documents or data proprietary to the company (what is often called "fine-tuning"), how to create commands or prompts that get the AI to do what they need it to do ("prompt engineering"), and how to evaluate the predictions or recommendations that the AI makes. Some of those skills are not directly related to the use of AI but are a consequence of deepening or upgrading work roles, such as learning to do competitive marketing analysis or scenario planning.

AI capabilities evolve quickly, requiring continuous learning. If the new reality engendered by working with AI is one of constant change, employee education must be a top priority. HealthCo, MarkCo, and UrbanGov all embraced the need for continuous employee reskilling. But each did so in different ways. For AI and data skills, HealthCo worked with the learning and development (L&D) team in HR to create an AI and cybersecurity "bootcamp" for employees. Some of the courses were taught by L&D members, some by university professors, and others by industry professionals and trainers contracted by HealthCo. The ethics and compliance division shared the cost of the bootcamp with HR, and employees could repeat the course each time the content changed.

MarkCo contracted with a local university to create custom programs to teach employees new skills in data science and AI. The university devised a set of tests that each employee needed to pass to certify that they were "AI ready." Every year, the test changed based on technological evolutions, and if employees did not pass the new test, they had to retake the program. With a much smaller budget, UrbanGov bought subscriptions for short courses from companies like LinkedIn and Udacity focused on AI, simulation, and data management. They crowdsourced a list of courses from these platforms that employees and managers found useful and encouraged each employee involved with the planning process, whether or not their work directly touched AI, to complete one course each month.

Companies that successfully supported learning had two things in common. First, they embedded the ethos of learning into their culture. Leaders and managers framed AI as a learning opportunity. Employees were not expected to know how to use AI perfectly upon adoption or immediately know how to segment their tasks around it. Instead, they were expected to explore its capabilities and determine how best to incorporate these new technologies into their work roles. Second, they provided time for employees to engage in the learning opportunities provided. HealthCo expected employees to devote at least two days per quarter to attending the bootcamp or refreshing skills learned in it. UrbanGov earmarked three hours per week for employees to take the online courses. Importantly, leaders at the companies periodically checked if AI tool evolution and learning necessitated revisiting segmentation and transition. Employees, knowing their skills would directly impact their roles, were highly motivated to learn.

Performance Evaluation

The final activity in the STEP framework requires managers to rethink the typical way they evaluate employee performance. Leaders in the companies that helped their employees use AI effectively shifted their thinking and practice around employee performance. Rather than treating productivity as an outcome upon which to evaluate employees, they reoriented to use it as an input. During the segmentation process, leaders put the onus on employees to determine whether they could use the tool in ways that made their work faster or more accurate. Thus, productivity became something employees themselves managed rather than something they were evaluated on. As a senior leader at MarkCo commented, "We used to measure how productive our employees were. That doesn't make sense with AI. Now we're trusting them to figure out how to use the AI in the most productive way. What we expect them to do and how we evaluate them has changed."

The major change in evaluation across all ten companies focused on determining how well employees were helping one another. Because people don't work in isolation, any changes that AI brings to one employee's role are likely to cascade into their interactions with others, affecting the work of many. Recognizing this, leaders who successfully integrated AI saw helping one another to learn and adapt as a crucial way employees could add value. As the managing director at UrbanGov observed, "The thing we need our employees doing in this age of AI is to help each other to learn and reimagine their jobs. It's a collaborative effort."

Consequently, the practice of performance evaluation shifted at all companies I studied in several ways. First, because of the segmentation and transition activities that were being done, the expectations for what tasks employees should do, and how, changed quickly. Annual performance evaluations no longer cut it because work roles changed multiple times in a year, often making objectives identified at the start of a performance period no longer relevant by the end of that period. Every firm in the study switched to a shorter performance evaluation period, most preferring to have quarterly meetings. Second, because

AI use was changing the tasks employees did, and hence their roles, they were constantly interacting with new people. Thus, a significant portion of the performance evaluation involved identifying the people they interacted with the most in that period and assessing whether the focal employee was a useful collaborator for them. In most companies, collaborators provided the evaluation of how well the focal employee helped them in their roles. The advantage of having a collaborator conduct this task was that they could evaluate the level of help they received better than a manager could. And because these evaluations were happening at much shorter intervals, the feedback an employee received about how they supported collaborators was immediately actionable; they could change their behavior to provide more help and assistance if necessary.

HealthCo's revamped performance evaluation system was the most aggressive and technologically advanced. Data scientists in HealthCo's HR division created a dashboard that drew data from employees' email communications, Slack use, and calendars to create social networks showing who they were most dependent on and who depended on them. Every six weeks, the dashboard sent a list of employees' most frequent collaborators, asking them to rate their interactions. The data were compiled and shared with employees and their managers, allowing them to gauge and improve their collaborative performance.

After two years of this new approach, HealthCo employees reported a 72 percent increase in satisfaction with the evaluation system. As one senior leader in the ethics and compliance division recounted, "The new system means that employees aren't just hearing from their managers; they are getting direct feedback from the people who depend on them. That is so important now that AI is changing how we work so quickly. We need everyone helping each other, and when our employees see themselves helping or find opportunities to help better, they do better."

The experiences of MarkCo, HealthCo, and UrbanGov show that the STEP framework can significantly benefit organizations. Mark-Co's use of AI to handle routine marketing tasks allowed employees to focus on high-value activities like competitor analysis and campaign testing. By deepening their roles, employees became more engaged and added more value to the firm. HealthCo's segmentation and transition processes ensured that employees were not threatened by automation but saw it as an opportunity to enhance their roles. By involving employees in the segmentation process and supporting their continuous education, HealthCo created a culture of trust and learning. UrbanGov's approach to upgrading roles demonstrated the importance of creativity in reimagining work. By reallocating tasks and responsibilities, they ensured that both junior and senior planners had meaningful, challenging work that utilized their skills and encouraged growth.

Many of the problems that lead to digital exhaustion in the workplace are systemic and can only be addressed by people with the authority to make change. Although your employees can deploy the eight rules in this book to be more proactive in dealing with their own digital exhaustion, the biggest benefits will come when you can create an environment that prevents many of the negative practices that create the conditions for digital exhaustion.

CHAPTER 5

Put On Your Own Mask Before Helping Others: A Guide for Parents

This chapter is NOT about how to parent. It is also not about helping your children succeed in the digital world. I won't give advice on whether you should limit your children's screen time, wait until they're a certain age to buy them a phone, or check their social media content periodically. I am not a child psychologist, a family therapist, or scholar of child and adolescent media consumption. There are plenty of books that do focus on these issues and plenty of experts with the right credentials for you to consult. If you're interested in where to begin, you can look to the notes in the back of this book to point you toward ones that I, as a parent and social scientist used to discerning what counts as good evidence and what doesn't, have found to be particularly insightful.

This *is* a chapter about how parents in the digital age can keep their wits about them. Here are some voices of parents I've talked to over the years who were clearly at their wit's end:

"I never thought being a parent would be this exhausting. I'm talking about being exhausted by the whole technology part of it. It's like, the carpools and the messages from school, and emails from their counselors, and the apps that tell me what field their games are on. And then I'm worried about whether they should have a phone and when is the right time for them to be on social media. And if they're going to do bad things online. It's too much."

—Steve, parent of two kids,
ages fourteen and eight

"I thought I was tired when they were babies and toddlers and they needed me physically all the time. Now I'm more tired because I've got to manage my kids' schedules all the time. Like right now, I've got two parents texting me asking if my one daughter wants a playdate. I've got the ParentSquare app I've got to keep track of because my son's class needs snacks for their party. Then I've got [medical app] where I'm waiting to see if the advice nurse is going to get back to me about my son's sore throat. How am I supposed to keep track of it all?"

—Amanda, parent of three kids,
ages thirteen, eight, and four

"My son's supposed to text me if he's running late from school, so I've got to keep my phone on me. But he doesn't text, so I go on Snap to see if he'll answer there since he's on there all the time. But then I get a DM on Facebook from a parent on my other son's flag football team asking me what time practice is, 'cause it always changes. But I don't know, so then I've got to look at the GameChanger app and I can't even remember if the coach sent a text about the game time or it was on the app or whether he called me and told me. I'm going crazy."

—Joaquin, parent of two kids,
ages seventeen and twelve

"I'm just really pissed. Our parents didn't have to deal with all this technology. Other parents weren't constantly texting them because there were no texts, and they couldn't see our grades on every assignment because there wasn't Canvas. Then Amelia's singing teacher lets the class out early because everybody has a phone and she just expects they can call home and get a ride whenever they need to. I had to wait until the class was over when I was a kid and my mom was ready to pick me up. I'm totally burnt out by all this technology. Our parents just faced most of the same issues that their parents faced. They didn't have all this technology. It's just unrealistic now."

—Rodda (my wife), parent of three kids,
ages thirteen, twelve, and ten

If you're like me, maybe just hearing that other people experience the same frustrations you do about parenting in this digital world will make you feel less crazy and not so alone. Of the hundreds of people I've talked with about parenting today, not one has told me that they think the digital technologies that inhabit our lives have made parenting easier. Sure, many extolled the virtues of being able to quickly make plans or to get a hold of their kids if they needed to, but the general consensus was that parenting in the digital age is difficult and exhausting.

We already know that the pressure of parenting kids who are themselves using digital technologies is a major source of stress and fatigue. There is a good deal of research that can instruct us about how to manage our kids' technology use. But it is surprising to me that there is so little research about how to deal with the exhaustion parents feel as the result of being trapped in the web of digital technologies, data, and communication that surround us and our kids. I submit that if we don't deal with our own digital exhaustion as parents, it will affect our children in negative ways. A robust line of research shows that high levels of parental exhaustion are correlated with poor parent-child relationships and, in the most extreme cases, contribute to neglectful and violent behaviors toward children. As flight attendants remind us every time we board an airplane, if the cabin loses pressure, we should always put on our own oxygen mask before assisting others, lest the lack of oxygen render us unconscious before we can help. If parents don't address their own digital exhaustion, they may slip into the terrible position of not having the time, strength, patience, or relational capital to help kids with theirs.

In this chapter, we'll explore various ways to reorient to our digital technologies as we parent. Because there are few published studies about this topic, I'll mostly draw from my own data and point to examples that have worked for the parents I've talked to. Each of these strategies embodies several of the rules we've already discussed, but they are packaged in a way that provides parents with a straightforward guide they can follow to reduce their digital exhaustion.

CALCULATE THE SHADOW HOURS

Mari is a mom of three kids, ages fourteen, eleven, and eight who lives in a suburb of Los Angeles. She works as a sales manager for a technology company, but her hours are flexible and she can choose whether to work from the office or from home on most days. Mari feels like this stage of her life is defined by driving. "Basically, I feel like I spend all day in the car," she told me. "My kids go to two different schools that each have different start times. My partner and I coordinate to drop them off, then sometimes I go to work or I head back home. It seems like at least twice a week someone's got a doctor's appointment or needs to get their braces adjusted, or something. So I'm picking them up from school and driving them there. Sometimes that takes like half an hour to get to one, depending on traffic." All three of Mari's kids play club soccer, year-round. Her son is dyslexic and sees a therapist once a week. Her oldest daughter is in the school's musical theater program, which often has practices and performances in the evening. Mari's partner, Nick, shares the work of driving the kids to and fro, but he travels a lot for work, so the bulk of the labor falls on Mari.

Mari's story is common among the parents I've interviewed who work in knowledge jobs and have kids between kindergarten and high school, though there are certainly regional and national differences in their experiences. Parents who raise their kids in city centers report slightly less driving because there are better public transportation options available. Parents outside the US report that their children participate in slightly fewer after-school activities than their overbooked American counterparts do. And parents in the rural US and countries in the Global South generally report receiving more assistance moving their kids around from extended family members such as parents, aunts and uncles, and siblings. Mari's situation—driving all over LA to move three kids here and there, largely by herself—might be somewhat extreme. But it is not rare. And despite these regional differences, parents around the globe seem to report roughly similar levels of exhaustion in aggregate. It seems that in terms of shuttling our kids around, there is a Parkinson's law for parental exhaustion: Exhaustion expands to match the number of times we commit ourselves to pick up and drop off our children at various activities.

Many parents work to make their lives as a chauffeur easier by coordinating with other parents. As Mari describes, "I think I'd go crazy and maybe I couldn't actually even do it all if I didn't get help. So, I've got my kids in several carpools, and I've got a group of friends who help each other out when we need it." That means that Mari is on text chains with other parents she knows personally and messages parents she knows less well through a variety of team, school-related, and other extracurricular-focused apps. "There is so much technology I use every day to try to keep track of it all and coordinate with everyone. It's really exhausting," she says.

Digital technologies make rapid coordination possible around very specific or time-sensitive issues, like letting a friend know that you need to adjust the pickup time for a kid when her dentist visit runs late. But using them in this way also creates significant "shadow labor." I first learned of the concept of shadow labor when I worked on a project for a large automotive company that was offshoring digital modeling work to India. That company had maintained engineering facilities and assembly plants in different countries for decades. But with the rise of the internet and the move toward digital product design, it became possible to offshore small tasks that might be completed in hours to workers in another country, rather than the design and manufacturing of an entire vehicle, which would take years. The American engineers loved the idea of sending tasks to India because they thought it would free them up to do higher-value, more interesting work. But they quickly found that although the Indian engineers were very competent, it still took a tremendous amount of time to assemble all of the files, write or call the engineer to describe what needed to be done, follow up with examples of prior work that they could use as a basis for the new work, and answer questions. I examined project tracking logs to determine how much time American engineers spent communicating about the work they sent to India. They estimated that an average-sized project should take an Indian engineer about five hours to complete. The project tracking logs showed that when the offshoring arrangement was new, American engineers were spending, on average, 4.5 hours communicating with the Indian engineers about the project. That means only half an hour was saved by offshoring the work to India. Fortunately, I was able to work with the teams to get the number of shadow hours down significantly over time. But they never went away completely. The dream of a "seamless transfer" or "24-hour engineering," as the company's senior executives touted, was never fully realized.

I never thought that the amount of shadow labor I talked about with Mari and other parents would add up to *hours* like it did for the automotive engineers. But I asked a group of ten parents to track the amount of time they spent communicating with their digital tools in order to coordinate with their spouses, childcare providers, and other parents to help move their kids around to various activities. Across the ten parents, the average came to 3.25 hours per week. When I dug in further to some specific instances, it turned out that—much like those early days of offshoring to India—parents were spending almost as much time coordinating rides for specific activities as it would take them to just drive their kid themselves. As Greta, one of the parents who did this exercise, explained:

> I coordinate with three other parents to do a drop-off for band practice. We try to have set days where one person drops off and another picks up, but it hardly ever works. We all have other kids and different schedules and work, so things just get messed up. So, look [she shows me a text chain on her phone that has thirty-two messages in it for one day], there is all this back-and-forth about how someone can't do pickup, can someone else cover, who is going to tell the kids, is practice actually ending at the same time today? Between looking at all this and responding and then having to go to the school app and my email to see when band practice actually ends, I spent about fifteen minutes today just coordinating the drop-off. It's only a twenty-minute round trip from my house. It would have practically been faster to do it myself.

But unlike for the automaker I worked with, there wasn't much hope that these shadow hours would decline over time. In the world of parenting, unexpected events that derail plans are to be expected. They are the rule, rather than the exception, which means that parents who use digital tools to help them coordinate will always be stuck in a situation where the ease of communication enabled by those tools will compel one parent in the group to ask for last-minute accommodations, which will then increase the shadow labor required by everyone. This was the same phenomenon that Christine Beckman and Melissa Mazmanian (whose research we learned about in chapter 1) observed among the working parents in their study. The same technologies that allowed people to outsource parts of their life also trapped them in a pattern of coordination with help-givers that added significant stress and exhaustion.

After realizing how many shadow hours she was spending to coordinate with other parents through her digital tools, Mari decided to take back three of her weekly chauffeur duties. As she described: "When I did that cost-benefit trade-off, it just became clear that I wasn't actually saving much time at all after doing so much coordination with everyone." As a result of her analysis, Mari decided to drop out of her carpool and take charge of driving her kids by herself. "I maybe spend like fifteen extra minutes now total," she said, "but not having to communicate with all those people and be constantly worried about if I was missing anything and not knowing what app to check or whatever makes me feel less fried. It's been totally worth it." What's most striking to me is that a few extra minutes in the car per week can feel less exhausting than having to keep track of all the communication necessary to spend less time in the car. Of course, not every parent has the luxury of simply choosing to do the work themselves. But Mari's example shows that we don't often think about the shadow hours involved in using the tools that are supposed to make our lives easier. Calculating the shadow hours demanded by your digital tools and determining if they are worth the cost paid by your exhaustion is one concrete step you can take to make smart decisions about how you play your role as a parent.

ON THE JOYS OF MISSING OUT

A lot of people I've talked with about digital exhaustion over the years use the cliché "drinking from a fire hose" to describe how they feel about their technology use. Our phones, tablets, computers, TVs, video game consoles, and smart speakers connect us to apps through which we have access to peerless amounts of data. But those data don't just sit there waiting for us to find them—the apps push those data at us all the time through messages, alerts, notifications, and more. As authors such as Nir Eyal, Adam Alter, and BJ Fogg have shown, many of the digital applications we use are purposefully designed to dole out data and information at unexpected intervals to keep us hooked. They are based on the psychological principle of "variable ratio reinforcement," which suggests that rewards given at random or varying intervals are more effective at maintaining a behavior than rewards given at fixed intervals. That means we're not just drinking from any old fire hose. Yes, the data comes out in a torrent when it comes. But it soaks us at a cadence that we cannot predict. That's partly why our phones and other digital devices are irresistible. We know something might be coming at any moment, so we might as well just check.

As most parents know, that fire hose usually soaks us at inconvenient times. We get messages from our kids' school while we're at work, and we get a text about the change in starting time for our kids' basketball game while we're making dinner. As we learned in chapter 1, these intermittent messages create context shifting across domains and arenas that can wear us out. We also learned in chapter 3 that data coming to us at random intervals also enhances our fear that we are missing out on some important discussion or activity. That FOMO is a real problem for parents, as Nikki, the mother of two teenagers, described: "We were having this issue on my son's club team that a bunch of the families were thinking of switching to a different team. There was all this back-and-forth on text, and then on the team app and phone calls. It was just going on for like a week. I was checking all the time, and my son told me I was obsessed. But I wanted to see what people were saying about it and I didn't want to feel left out." Although Nikki knew that all of the discussion didn't need to be attended to in the moment, she was, as she described, "Like a moth to a flame." She feared missing out and her FOMO encouraged her to let the many notifications disrupt her other activities.

There is always some issue happening or some drama unfolding in real time on our devices, whether among our friend groups, with the parents of our kids' friends, at school, in a sports team, in a music club, or wherever. And the fear of missing out on this kind of important information that pertains to our kids' lives, and our own lives by extension, has been shown to be a significant driver of "parental burnout" writ large, and of digital exhaustion more specifically. How do we keep these fears at bay? A tech company mistake may contain one answer.

On October 4, 2021, Meta, the parent company of Facebook, Instagram, Messenger, and WhatsApp, experienced a severe technical failure. For nearly seven hours, users were unable to log on to Meta's products, preventing billions of users from accessing their accounts or utilizing any resources typically available on these platforms. Additionally, communication through Messenger and WhatsApp, two primary global messaging services, was disrupted. In the immediate wake of this outage, Tal Eitan and Tali Gazit of Bar-Ilan University surveyed more than five hundred Israeli adults who were frequent users of Meta's products, measuring, among other things, their levels of FOMO, and asked them to respond to open-ended questions about their experience during the outage. As expected, they found that people initially experienced high levels of FOMO right after the outage began. They worried that they were missing out on information and activities that others were sharing on the platform while they were stuck in the dark. But the study participants noted that as they began to realize the outage was global, they felt a sense of relief. They didn't feel like they were missing out because no one else could access the platforms either. As the authors described, many participants expressed a feeling of joy.

Intrigued by these findings, the authors launched a more focused and structured study to explore whether people could experience the joy of missing out (JOMO) in situations in which they knew they were purposefully choosing to disengage from social media while everyone else around the world was still active on it. The follow-up study suggested that it was difficult for people to trade FOMO for JOMO. The findings showed that high school– and college-aged individuals who were single, had weaker psychological well-being, and used social media frequently, did not typically experience JOMO when disconnected from social media. They were fearful and anxious about being left out. However, people in their prime child-rearing years who had strong psychological well-being scores were able to experience JOMO after an initial period of withdrawal. Other studies that have examined how people can experience JOMO suggest that using digital devices mindfully and making an intentional decision to walk away from the fire hose can lead to a blissful feeling associated with not knowing all the details of what is going on at a given moment. Those studies also show that finding non-technological activities that are cognitively engrossing makes it easier to care less about what is happening on your apps while you're away.

When Nikki learned about the idea of JOMO, she was initially doubtful. "I mean, that sounds weird," she told me. "But I guess I could see how you could maybe start to feel sort of Zen if you knew there was a bunch of stuff swirling around you that the other parents were dealing with and you made a choice not to be involved in it." She committed to shifting her behavior to give it a try. She turned off the push notifications on the various apps she used to coordinate her kids' schedules and committed to only looking at them at the end of each day. Further, she decided to not read and respond to the flurry of messages that were being exchanged on the group text with the parents of her son's club lacrosse team until the end of the day as well. After two weeks of trying this new approach, Nikki reported some success: "It's kinda freeing knowing that all the rest of the parents are scurrying around and I don't have to. In a sick way, I kind of got to the point of where I heard my phone ding and thought, 'I don't have to worry about what's going on till later.' That actually has been good. I do feel a little more energized." The trade-off, of course, was that sometimes decisions got made without Nikki's input because she wasn't quick enough to respond. But by and large, she reported that missing out on giving some suggestions caused her much less stress than tracking the conversation in real time.

JOMO only happens if you're willing to make an attitude shift and decide that it's worth sacrificing a bit of control to avoid worrying about what everyone else is thinking or doing. Most parents who have told me they've tried this approach report that the things they miss out on typically end up being trivial and they come to relish showing up later to the conversation. As Marco, a father of four, told me, "I really like how now I check my phone and see a bunch of stuff that happened between all the other parents and I think, 'Those suckers wasted all this time figuring it out while I was playing basketball with my son.'"

BE LESS CONNECTED TO FEEL
MORE CONNECTION

In scholarship on social networks, a popular way to measure tie strength—a proxy for how robust a relationship is between two people—is to ask a survey respondent to select a name from a roster of people they communicate with and then answer two questions. The first question asks the respondent to rate how frequently they interact with the selected person. The second question asks some variant of how "close" the respondent feels to that person. In most areas of social life, people's ratings of frequency of interaction and closeness are highly correlated. That is, if you communicate often with someone, you also tend to rate them as a close contact, and vice versa. When two people assign high ratings of both frequency and closeness to each other, network scientists say those two people have a "strong tie." And when they assign low ratings on both measures, that indicates a "weak tie." When I used to teach a networks class to business school students, I joked that a weak tie is someone who would help you carry a sofa, while a strong tie was someone who would help you carry a dead body. Occasionally, I would get a laugh.

Although there are many studies of parent-child networks, and parent-teacher networks, and the social networks of parents broadly (which typically include other adults who are parent friends, but who are not necessarily parents of their kids' friends), there are almost no empirical studies of networks of parents coordinating their lives together in the digital era. So in what follows, I will rely mostly on qualitative insights. As I've learned more about the ways that parents use digital technologies to communicate and interact with one another, it has become obvious that among networks of parents, communication frequency and emotional closeness are not often correlated as they are in other aspects of people's lives. Corinne provides a common example. She's the mother of two daughters, and she estimates that on a weekly basis she uses her digital tools to interact with roughly fifteen other parents. "I only consider one or two of those people my friends," she says. "Most of them are either parents of my kids' friends, or parents of kids that are in my kids' classes or that do other activities with them. But, you know, you're all on the same apps and you see them on Facebook and Instagram and you're texting with them about all kinds of stuff. So they're definitely part of your network." When I asked Corrine to do a similar analysis to those run by network scientists to determine who in her network of parents she rates high on both communication frequency and closeness, she responded emphatically: "Pretty much no one." She went on, "I communicate with most of these people A LOT, but I don't really know most of them. I'm not close with them. I mean, with some of them I don't even know their last names or how many kids they have. But I communicate with them way more than with my own friends or, like, my siblings." Like most parents I've talked with, Corinne's network of parent relations is filled with would-be couch movers but is noticeably light on people who would help her dispose of a dead body.

To understand why weak ties in the network topology of parenthood are so important to consider in regards to digital exhaustion, let's revisit Rule #2 (Make a Match). When we discussed how to choose the best digital tools for the job, we learned to be careful not to undermatch. If we choose too lean of a medium for our coordination need, we can get caught in an endless cycle of confusion and inference, which creates the demand for even more communication. If we overmatch by choosing too rich a medium for the job, we can end up wasting our time and the time of others, which can also lead to exhaustion. Almost all of the communications and interactions between the parents I've studied happen through their digital tools. A quick text here, a like on Facebook there. You get it. Most of the time, parents are implicitly following Rule #2 by choosing lean media because they recognize they don't need to hop on a Zoom call or drive to someone's house for a face-to-face conversation to simply sort out whether it's their turn to pick up little Bobby. But the problem with applying Rule #2 exclusively in this context is that many of our interactions with other parents lack the multidimensionality of our relationships with our real friends, family, and coworkers. There aren't many natural opportunities beyond the text to build the kind of relationship that would require us to make a different kind of match. As Corinne wisely observed, "If I wanted to, I could just be on text forever with most other parents and never actually talk to them or get to know them because all we do is decide who goes where."

The problem with this setup is that we can be very connected to other parents in our daily lives through our digital technologies but not feel a connection with them. In network terms, we can have high communication frequency without emotional closeness. In a recent study led by Jeffrey Hall at the University of Kansas, a research team asked 429 people to identify interactions with network partners in which they expended a lot of energy and felt depleted. The findings showed that feeling disconnected from someone during a particular interaction was associated with high energy expenditure. It took a lot out of people to communicate with those they didn't feel connected to. If someone felt disconnected and lonely after an energy-intensive interaction, their subsequent interactions were more likely to also be energy depleting. And to make matters worse, the authors concluded that, "If people are already exhausted from energy-intensive interactions and still choose to be around others, they must expend more energy, resulting in greater fatigue." A vicious cycle for sure.

Several parents who scored low on the exhaustion scale discussed how they get out of this vicious cycle. The common theme across their stories was that they reduced the frequency of communications with certain people using lean media and switched to using richer media with them less frequently. Garth explained it to me this way: "One thing I decided to do was rather than text about each little thing, I would call them at the beginning of the week or ask them to meet me after school on the playground to discuss the whole week's schedule. I actually really liked that more, because then you kind of get to know them better and talk about stuff that's not just your kids." Rather than orient to each individual pickup and drop-off as a discrete coordination need, Garth refocused to think about the entire week. When he looked at it this way, the week was a complex coordination problem that demanded a richer medium to sort out. Using that richer medium (here, phone or in-person interaction) made Garth feel closer to the other parents. And, as he described, that increased closeness helped to defray feelings of exhaustion: "When I actually connect and get to know them as a person rather than a disembodied number on my screen, I walk away from that conversation feeling more energized and not exhausted like I do after sending thirty texts." Garth's strategy to be less connected to feel more connection also helped reduce the number of shadow hours he spent each week coordinating with other parents. Many other parents told me that meeting people in person or talking with them on the phone also blunted their tendency to make the assumptions that were so easy to fall into when they didn't really know the other parent.

Interacting with a "disembodied number" is not like those sporadic, lightweight interactions on social networks that energize as they give us the ability to learn vicariously. Frequent interactions with people without emotional closeness is energy draining. Finding other ways of interacting that make us feel more connected to other parents may seem laborious in the short run, but the parents who have successfully implemented this strategy say it is a worthwhile investment.

DITCH THE DEVICES IN FRONT OF YOUR KIDS

The perspective that kids have on their parents' use of digital technologies is one area that has actually received a fair amount of research attention. Kids of all ages seem to notice when parents are on their phones, and they don't like it. Numerous studies report that when parents are distracted by their digital tools, including phones, social media, and games, they tend to pay less attention to their kids. As a result, children tend to either withdraw or act out. As one example, a study of couples with children ages five and younger showed that the increased parental use of digital technologies was correlated with increases in behavioral issues in their children, as did another study of children between four and ten years old. Parents who experienced stress and exhaustion as the result of their caregiver role turned to a variety of digital technologies more frequently for an escape. But the more they turned to their digital tools, the worse their kids behaved, which just increased their stress and exhaustion. Similar results were uncovered in a study of mothers of children ages one to five; it found that maternal depression was associated with phone use that distracted mothers from their children, which in turn was associated with increased technology interference in parenting. Finally, to top things off, a systematic review of twenty-seven studies including more than 5,700 participants concluded that parents engaged with smartphones around children ranging in ages from one to eighteen were less verbally and nonverbally responsive to their child across different types of settings. These findings reinforce what most parents come to learn after just a short while on the job: Our kids are *always* watching us. And when our digital technologies distract us from them, they have a propensity to act out in negative ways.

It doesn't seem too surprising that children of all ages would be upset by anything that takes their parents' attention away from them.

So what's so special about digital technologies? Alma, a mother of two teenage boys, offered a plausible theory: "What I've noticed with my sons over the years is that my tablet or phone or whatever is like a signal. It signals to them that I'd rather be somewhere else or with someone else than them. If I'm reading a book or cooking or whatever, sure, I'm doing something else. But I'm not somewhere else. Does that make sense? They react to that. They always have, and they still do even though they're teenagers and on their own devices all the time." Just as we discussed in Rule #8 (Be Here, Not Elsewhere), our digital tools can teleport us to different places with ease, making it difficult to be here with our kids. I like Alma's use of the word "signal." If we choose to spend our time in front of our digital devices when we're around our kids, we inadvertently send them a signal that there is something more important on the other side of the screen than they are. Cate, a mother of one emotionally mature adolescent girl, teared up while telling me about the "serious" conversation her daughter asked to have with her: "She said, 'Mom, I want to tell you something and don't get mad. But I just wish you'd give me more attention. You're always on your phone. I don't know what you're doing, but it doesn't seem important, and it hurts my feelings that you don't want to hear about my day.' Then she told me that she was going to go to college soon and that we should do more things together while we still have time. I mean, I felt like a total jerk. I am doing nothing on my phone most of the time. She's right." If your kids, like Cate's or like mine, have ever given you a hard time for being on your phone too much, or spending too much time at your computer, or watching too much TV, you've probably felt that sharp realization that your attention does really matter. As my UCSB colleague Robin Nabi, an expert in kids' media use, notes, "The takeaway is for parents to be more mindful of how often they are using their phones around their children. Where their eyes are sends a message to their children about what's important."

"I would work from home, but I don't want my kids to see all the screen time I get."

Perhaps for this reason, a recent slate of studies has been showing that parents who spend significant time on digital devices in front of their children feel increased stress, anxiety, and guilt about their effectiveness as parents. One interview study conducted by researchers at the University of Michigan found that parents often felt guilty and angry that their own use of digital technologies at home teleported them away from their children and gave their children the idea that being on a device all the time was OK. As one of the few studies that has asked children directly for their interpretations of excessive parental digital technology use, the authors were able to report what kids actually thought about the matter. As the authors wrote:

> While all participants had rules around technology use, both parents and children reported breaking them. Children reported frequent instances of parents using phones during dinner time or other family times. Children, however, excused these violations if they perceived them to be work-related. Violet said:

"Because of [my dad's] schedule, he has a bunch of people that need to communicate with him. [His baseball players] always text him when they need to know times for practice, and his assistant coaches, and then all of his people from work. We joke that he's worse than we are." Similarly, Shawn reported that both his parents occasionally used their phones during family time. Shawn said this rule violation felt disrespectful: "I feel a bit disrespected . . . well, it depends what the call is for. Something random like a friend or something, I feel a bit ignored. But if it's for work, I completely understand."

It's interesting that even older teenage children were bothered by their parents' technology use, but that they justified it if the use was related to their jobs. On this point, though, Umberto, a father of two teenagers, told me that knowing that his kids thought that his use of digital technologies for work was acceptable made him feel even worse: "My kids tell me that if I need to use it [my phone] for work like during dinner or something, that it's OK. But honestly that makes me feel worse than anything. Not only am I disrupting our dinner to be on my phone but I'm showing them that work is more important than them and than our homelife. That makes me feel crappy and embarrassed. I've got to stop."

So if our digital technology use in front of our children is a source of our own exhaustion, the solution is simple, though not easy: Stop using your devices in front of your kids. The parents that I've interviewed who seem to do this best don't give up their digital tools completely, of course. Instead, they work hard to apply Rule #1 (Stop Using Half Your Tools), Rule #3 (Batch and Stream), Rule #4 (Wait), Rule #6 (Act with Intention), and Rule #8 (Be Here, Not Elsewhere.) For many parents, applying those rules means setting up specific times during the evenings and weekends when they will use their digital tools and refraining from using them at other times. Fatima, a mother of three kids ranging in age from eight to fifteen, described how she has created a "mommy homework hour" at the same time that her kids do their homework: "That's all the time I spend on my devices now in the evening. I wait and respond to everything then. I stopped using a bunch of apps and I'm more intentional about what I'm doing to fit it into that window." As she reports, "I'm pretty sure my kids notice the difference and it makes me feel better as a parent."

Based on these emerging data, it seems that staying off our devices as much as we can in front of our kids does double duty for reducing our digital exhaustion. First, it can stop our own feelings of guilt, anxiety, and anger that arise from feeling like we're not prioritizing our kids as we teleport to other locations in our minds. Second, it can teach our kids healthier habits about how and when to use their own digital devices, which can also help reduce the exhaustion we feel about being bad technology role models for our kids. As we've learned, it's not easy to resist the temptation of our tools. But applying the rules we've discussed throughout this book can help us to weaken the tractor beam enough to make our own decisions about whether the time is right to pick up the phone, log in, and lose ourselves amid the 1s and 0s.

CHAPTER 6

The Artifice of Intelligence:
Living and Working with AI

We are entering a brave new world in which digital technologies powered by artificial intelligence are going to play a central role. Conversational agents built on large language models (LLMs) are trained on unfathomable amounts of text scraped from all corners of the web. Linguists have argued for many years that human language evinces deep structural properties—underlying patterns that are common to all languages. Although the provenance of such deep structures is hotly debated, it is undeniable that the characteristics of our language are so ingrained in the way we write and think that with only the tools of data classification and statistics at their disposal, such technologies can predict what humans are likely to do or say with relatively high degrees of accuracy. Noam Chomsky, famed MIT linguistics professor and originator of the theory of universal grammar, is less impressed than I am by the capabilities of generative AI. Excoriating ChatGPT for being a "lumbering statistical engine for pattern matching, gorging on hundreds of terabytes of data," he and his coauthors write: "Their deepest flaw is the absence of the most critical capacity of any intelligence: to say not only what is the case, what was the case and what will be the case—that's description and prediction—but also what is not the case and what could and could not be the case. Those are the ingredients of explanation, the mark of true intelligence." These comments remind me of the witty words of French sociologist Jean Baudrillard in his book *The Transparency of Evil*: "Artificial intelligence is devoid of intelligence because it is devoid of artifice."

No matter your view on the question of intelligence, it's clear that these new technologies are formidable prediction machines. In many ways, they may come to know humans better than we know ourselves: Soon they will have read the whole of our combined written knowledge and extracted from it the deep patterns of thought and action that evince and reflect our humanity. Those same artifacts that teach each new successive generation how to be human are also teaching AI what it means to be human. Whether that thought sparks fear or delight in you is beside the point. AI-powered tools are learning—from the cultural products humans make. No one programmed ChatGPT to have (or mimic) theory of mind, just like no one programmed it to be able to translate one language into many others. Those capabilities have, as Stanford cognitive scientist Michal Kosinski says, "spontaneously emerged" as the algorithms that power LLMs make predictions based on the patterns they surface from the depths of human language. In the best cases, they tell you what people statistically similar to you would statistically like to hear based on the statistical similarities of the prompt you entered to the statistically significant patterns in the text it has parsed. In the worst cases, they just make shit up.

And therein lie AI's implications for digital exhaustion. Depending on how we use them, technologies powered by AI could either save us from technological overload or become the worst contributors to it.

I'm mindful of a meme circulating on social media that says that among the fastest things in the universe, the only thing surpassing the speed of light is the speed with which people are becoming experts in AI. So I won't prognosticate too much. Instead, I'll lay out three possibilities for how AI might affect our levels of digital exhaustion. As we'll discuss, the choices we make about how to use AI in each of these areas will be the critical determinants of whether those levels increase or decrease. Of course, we are going to be the creators of our future with AI. So let's explore how we can make the choices that create the future we want to live in.

BEWARE THE COMING DELUGE OF CONTENT

"You ain't seen nothing yet. That's the expression, right?" This was Karl's response when I asked him if he's been seeing a growth in AI-generated content. Karl is the chief AI scientist for a midsize SaaS company. His German-accented English is perfect—he knew that was the right expression and he used it with aplomb. One of the major patterns emerging in these early days of GenAI use is content creation. Mandy, a senior account sales rep at an electronics company in Dallas told me that she has at least tripled the amount of written content she's produced in the six months since she began using Anthropic's Claude at work: "It used to take me a week to write a good sales report. Now I can do at least three in a week. It's like that for just about everything. I've counted up how much stuff I've been able to do, and I'm probably producing three times as much stuff as I used to in the same amount of time." Dashun, a technical writer for a medical device company, offered similar estimates after six months of using ChatGPT: "I'm probably doing 30 to 40 percent more output now than I was before using ChatGPT. Mostly it helps me get the basics done so

I can do more of the complex stuff. It's great." Karl's company was producing an interactive FAQ interface for its customers that was powered by GenAI. When customers interacted with the interface, the machine learning model updated so it could produce new and more accurate FAQs in the future. With more FAQs out there to learn from, the AI could produce more FAQs more quickly. "We need data to build better content," Klaus told me. "It's a great flywheel when you can use AI to catalyze the production of more data that your models can learn from."

There have been a lot of studies looking to quantify the productivity increases attributable to GenAI in the knowledge workplace. One field experiment led by researchers at Harvard showed that consultants at the Boston Consulting Group who were "using AI were significantly more productive (they completed 12.2% more tasks on average, and completed tasks 25.1% more quickly), and produced significantly higher quality results (more than 40% higher quality compared to a control group)." Another survey by Microsoft of users of its Copilot tool revealed that 70 percent said they were more productive, with 73 percent of respondents saying that they could complete tasks, including content creation, faster, and 85 percent saying it helped them get to a good first draft of whatever they were working on faster. Another study conducted by researchers at OpenAI and the University of Pennsylvania used the O*NET database, which includes detailed work activities and tasks for more than one thousand occupations, combined with data from the US Bureau of Labor Statistics, to model which jobs could be sped up by GenAI. They concluded that "with access to an LLM, about 15% of all worker tasks in the US could be completed significantly faster at the same level of quality. When incorporating software and tooling built on top of LLMs, this share increases to between 47 and 56% of all tasks." Whether generated by experiments, surveys, or mathematical models, the available evidence points in one direction: People are doing things faster with AI. And in the world of content production, that means that they're producing more—much more—than they ever have. So much more, in fact, that a report produced by the European law enforcement group Europol warns that there is so much AI-generated content coming online every day that it will be nearly impossible for law enforcement experts, and the general public, to know if an AI tool generated content or if a person did. The report anticipates that by 2030 the vast majority of content on the web will be produced by AI. Does that qualify as a deluge?

A funny, and sobering, incident in one of my classes recently brought the coming deluge of AI-generated content into focus. A senior vice president of engineering at a very cool start-up agreed to speak to my students in the Master of Technology Management program at UCSB. He discussed how his company was deciding whether to build its own LLM models internally or to buy them from an AI company like OpenAI or Anthropic. After the session concluded, I asked students to write thank-you notes to our guest that emphasized what they learned from him and how it related to the content we were covering in the course. Sixteen students opted to submit thank-you notes. Here are snippets from several of them, showing the second and fourth paragraph of three different student submissions.

Examples of Similar Paragraphs in
Thank-You Notes Generated by GenAI Tools

(Notice the similarities in diction, syntax, and paragraph structure)

STUDENT 1

2nd Paragraph

One of the key takeaways from your talk was the concept of Software as a Service (SaaS) and its transformative impact on the real estate industry. Understanding how your company leverages cloud technology to streamline property management was particularly inspiring. This session was well executed; discussions centered on cloud computing and its role in enabling businesses to scale efficiently without the burdens of traditional IT infrastructure. Your explanation helped clarify my appreciation for this unique idea and showed how it can be a valuable tool in business optimization.

4th Paragraph

Inspired by your insights, I plan to incorporate more AI-driven solutions into my future projects, particularly focusing on automating repetitive tasks to improve efficiency. Additionally, I will prioritize data security and privacy, being more aware of how these frameworks can affect my industry. Your perspectives were helpful as I realized a need to focus more on the needs of my clients, which will undoubtedly shape my approach to future technological challenges.

STUDENT 2

2nd Paragraph

One of the key lessons I learned from your presentation was the transformative impact of digital intelligence on business processes. Your demonstration on how AI is revolutionizing real estate property management by reducing waste and enhancing decision-making was insightful and highly captivating. This idea clearly ties into other sessions on property management and technology management, which we have been learning about in the context of real estate disruption. Your example of your company's AI's integration set the foundation for subsequent lessons of the course, making this discussion appear to have added more insight than I had expected.

4th Paragraph

Inspired by your talk, I am now convinced to apply a systematic, data-driven approach to my work. Research on the feasibility of AI functionality for decision-making across multiple property management contexts would have been inconceivable even a few years back. Your talk has expanded my sense of how AI frameworks can revolutionize the way we approach complex issues.

STUDENT 3

2nd Paragraph

Another key takeaway was your discussion on the value of integrating automation and the challenges of AI deployment within real estate. Your emphasis on the changing technological landscape in the real estate industry inspired me to consider innovative business models and creative problem-solving approaches. Understanding the steps you have taken to streamline real property-related processes was crucial in helping me appreciate how to assess situations and apply critical thinking within a business context. Your practical approach to these topics, paired with your specific industry experiences, was incredibly valuable.

4th Paragraph

Motivated by your talk, I plan to take a more interdisciplinary approach to my education, focusing on identifying potential uses for AI in my work with decision-makers and clients. I am looking forward to furthering my understanding of how technological advancements can be applied in my field.

What struck me about these submissions was how similar they were in diction, tone, paragraph length, and structure. I didn't really have any worry that the students used a GenAI tool to do the entire assignment or to commit plagiarism; if you actually read these boring entries, you'd see that the content is distinct and matches what I expected them to get from the guest speaker's visit. Instead, I presumed what happened was that the students wrote up some brief notes then submitted them to a GenAI tool (probably ChatGPT) to expand the content. Obviously, I couldn't send sixteen nearly identically worded and structured thank you notes to our guest speaker—especially one who builds GenAI products for a living! So I sent an email to all the students who submitted the notes. I wrote that I could tell exactly which students had used GenAI to doctor their thank-you notes, and that if they wanted credit for the assignment those students would need to send me the original they had used as a model for their AI assistant. Remarkably, but not surprisingly, each student who I suspected used GenAI sent me the original version of their thank-you note. The students who I suspected hadn't used GenAI didn't send anything.

When I compared the AI-doctored thank-you notes to the original, non-AI-assisted drafts, I noticed two things. First, the originals were much more colloquially written and heartfelt. They were much better thank-you notes, at least from a cultural standpoint. Second, they were *much* shorter. I ran a comparison, and the average non-AI-assisted note was 193 words. The average AI-assisted note contained 592 words. The volume of words more than tripled with the use of AI. If my students were working in Mandy's electronics firm or Dashun's medical device company, they would have been right on track to produce three times as much content too. Since I thought he would appreciate the humor of the whole situation, I sent the guest speaker both sets of notes. Here was his email reply:

> OMG! It was so funny to read those. The original notes are much better and I appreciate that they really thought about the insights from our conversation. I was glad to see they resonated. If you had just sent the AI-doctored notes I would have put them all into ChatGPT and asked it to summarize them for me so I could see what themes the students picked up on. To be honest, I wouldn't have read through all of them, it would have been too tiring.

The students' solution to the thank-you note assignment was to expand their content using GenAI. Then, my speaker's solution was to use GenAI to shrink the content back down so he wouldn't get so exhausted going through it all. He's not alone in using this strategy. A joint study by Asana and Anthropic revealed that the top two uses cases for GenAI in the workplace are 1) email generation (37 percent of all respondents reported using it to elongate their emails), and 2) information summarization (34 percent say they use it to get short-

© marketoonist.com

ened summaries of content, including emails). The irony here is impressive. In order to "save time" and "be productive," someone takes their notes or brief outline and puts it into an LLM to expand it. Then they send it to someone who uses an LLM to summarize it. We've taken a chain that should be three steps (write > send > read) and made it five steps (write > expand > send > summarize > read). The most efficient thing to do would be to just cut the pretext of feeling like you need to send something detailed and send the original short text, since that is all the person on the other end wants to read anyway.

Justin, a senior leader at an agriculture science company, told me that he suspects that nearly 70 percent of the internal company documents he's seen since the start of 2024 have been amplified and lengthened by AI tools. Jimena, a manager at a workflow automation company, puts that number closer to 80 percent. Both Justin and Jimena used the same word when I asked them how the proliferation of AI generated content makes them feel: "Exhausted." It's sure fun to play with GenAI and make it turn our drivel into eloquence or watch it expand a few rough thoughts into a fairly cohesive narrative. But we need to ask whether that is really the best use of AI. How much more content do we really need? Is the amplification and reduction cycle really how we want to spend our time and energy? These are critical questions for how we think about the future of AI.

My recommendation is that we use generative AI as much as we can to reduce rather than to expand. That's going to be difficult for a lot of people. My students were enamored with how it could make their language sound more "professional" and expand their points to make it look like they thought through an issue deeply. Others marvel at how easily GenAI tools create text or images and want to show off these new creations. But when we set ourselves up to produce less, not only is it less exhausting for us, it's also less exhausting for the people with whom we interact. Using the principles that undergird these rules in conjunction with AI should help us figure out how to reduce. That same Microsoft study that suggested Copilot users were 70 percent more productive did find some glimmers of hope about reduction. The research team asked a sixty-two-person blind panel to rate the clarity and conciseness of multiple email messages, some written with the help of Copilot and some without. They reported that the average email written with the help of GenAI was 19 percent more concise and 18 percent clearer than emails written without it. If we push to continually reduce rather than add content—a countertendency to be sure—AI may indeed be of tremendous help in decreasing digital exhaustion.

EMBRACE ARTIFICIAL SERENDIPITY

The main reason I was able to detect that the students from my class used AI to doctor their thank-you notes was that I know their actual writing. Most of it is fine, but not great. Their sentence structure is clunky, punctuation is spotty, diction is boring, and clauses—don't get me started on the clauses. But that's typically OK in this kind of class, where my goal is to help them learn frameworks for creating and leading digital businesses. It's just a bonus if I can help them with their prose. But those AI-adjusted notes that were submitted didn't read like their normal writing. They were too "polished" in terms of sentence structure and punctuation. They used words that I've never once heard come out of these students' mouths. There's a reason for that. In a brilliant piece of investigative reporting, Kevin Schaul, Szu Yu Chen, and Nitasha Tiku of *The Washington Post* worked with researchers at the Allen Institute for Artificial Intelligence to analyze Google's C4 dataset, which contains the contents of over fifteen million websites that have been used to train Google's Gemini and Meta's LLaMA generative AI tools, among others. Their analysis, which is certainly worth reading, reveals that the training dataset was disproportionately populated by websites from industries including journalism, entertainment, software development, and medicine. The three biggest sites harvested for content were 1) patents.google.com, which contains text from patents issued around the world; 2) wikipedia.org, the free online encyclopedia; and 3) scribd.com, a subscription-only digital library. Unfortunately for book authors like me, also high on the list were sites like b-ok.org, which is an online marketplace for pirated books that was recently seized by the US Department of Justice. As you might imagine, the writing on sites like these doesn't mirror the parlance of the masses. It's typically much more erudite and somewhat stuffy, and it traffics in complex ideas.

But if that kind of writing is what LLMs have been trained on, that's the kind of writing they are going to mimic. When Camille Endacott and I studied users of the AI scheduling agent Lisa, we had access to all the emails that were sent by the AI agent, and we could see that the diction, sentence structure, and other characteristics of writing looked similar no matter who their principal was. The emails didn't reflect the individual style of the principal using the AI agent;

rather, they looked similar to what my students wrote. Even more startling was that Lisa followed similar scheduling patterns for different principals: It had certain preferences about when to start meetings and how much time to give between them. When we asked Lisa's developers about this similarity in language and scheduling patterns, they were unabashed. They told us that they'd chosen to train Lisa on data from people they considered expert schedulers, and that in the reinforcement learning phase of training they rewarded the machine learning algorithm to favor the patterns of these "master schedulers" over other patterns—like those that would come directly from the principal's own past scheduling behavior. As the senior data scientist at the company told us, "People are notoriously bad at knowing what they want and they are unrealistic when they schedule. So we have the AI learn from people who are master schedulers instead."

The long-term concern with training AI agents on similar datasets is that they will likely produce similar decisions in different contexts. And as more people act on those decisions, they will generate patterns of behavior that are similar and leave similar digital records from which LLMs will continue to learn, thus generating an increasingly smaller and homogeneous set of decisions. An infinite loop, as it were. In our study context, that might mean that if people agree to schedule following Lisa's recommendations, their calendars will start to look similar to the calendars of other people. In our data, that actually looked like less time between meetings, meetings that never started before 9 a.m., and meetings that were clustered in the morning. Over time, as the AI expands to really consider these calendar data as learning inputs, it will develop patterns based on them and recommend scheduling behaviors that stomp out any variation between people at all. Maybe similar scheduling structures for all is not a big deal in the grand scheme of things. But what about clothing recommendations or advice on how to choose your next career move?

Clearly there are areas in which we don't want our AIs to act homogeneously, nor do we want to become lemmings.

Innovation scholars have long shown that new ideas often result when people connect seemingly disparate and unrelated ideas. They call this "recombination." The evidence is pretty clear that people who excel at generating new ideas that are useful and valuable tend to connect social groups that aren't themselves well connected. As the sociologist Ronald Burt describes, these innovators have a "vision advantage" because they can see the problems in one group and connect them to solutions from another. These innovators develop their vision advantage because they live in different social worlds at the same time. They might be a Wall Street analyst by day and a bassist in a punk band by night. Or, as is likely more typical, they work in product development at their company but they spend a lot of time developing relationships with coworkers in marketing. Burt's research has shown that people who occupy these positions as brokers across groups are markedly more innovative than their colleagues who live in just one social world, and that they receive significantly higher raises and faster promotions because of it. Although there has been a lot of research into the network structure of innovation, the jury is still out on whether you can successfully train someone to get into these unique brokerage positions and, if you can, whether they can learn to seize the opportunity for recombination when it arises. The most convincing research I've seen shows that there is a lot of serendipity involved in the whole thing. People end up spanning social groups for so many reasons—the skills they developed in past jobs, where they were assigned to sit in the office, who they were friends with in college—many of which just come down to luck. And their ability to recognize that two ideas might go together seems to have a lot to do with whether they are in the right place at the right time and have the requisite knowledge to identify and understand what is happening in front of them. The one thing that is often in their control is whether they choose to do the work to remain members of multiple communities simultaneously. But people who span different social worlds, especially in the workplace, get burned out fast because of the social and cognitive difficulties involved with moving back and forth constantly across cultures. Consequently, they don't typically last as brokers for too long, and their ability to make the connections necessary for innovation is often short-lived.

As Ethan Mollick, author of *Co-Intelligence*, suggests, LLMs may be able to step in when humans are too exhausted to continue because they are "connection machines." As he writes, "They are trained by generating relationships between tokens that may seem unrelated to humans but represent some deeper meaning. Add in the randomness that comes with AI output, and you have a powerful tool for innovation. The AI seeks to generate the next word in a sequence by finding the next likely token, no matter how weird the previous words were. So it should be no surprise that AI can come up with novel concepts with ease." That makes sense to me. But such a suggestion is predicated on the assumption that the LLM has access to a wide and diverse set of training data. If the coming deluge of AI-generated content— content produced by AI in response to patterns it has identified in human-generated content—produces more behaviors and new content that drives homogeneity, the diversity of options AI will have to draw from will winnow over time. Shi Feng and James Evans of the University of Chicago conducted one of the largest analyses of research papers and patents ever undertaken. Their findings showed that "surprise in terms of unexpected combinations of contents and contexts predicts outsized impact" in scientific discovery. As they conclude, "surprising advances emerge across, rather than within researchers or teams—most commonly when scientists from one field publish problem-solving results to an audience from a distant field." We need diverse and different sets of knowledge to come into contact for innovation to occur. It seems entirely possible that by training AI on its own outputs, we could end up killing those important opportunities for serendipity so essential for innovation.

Another way of thinking about this is to remember that LLMs capture patterns through repetition, essentially encoding what they see most often. This means they tend to average out the data they process; they hold on to the average pattern they have identified, not the outliers. This is likely why research shows that use of LLMs significantly improves the work of poor performers, bringing their quality up to average or slightly above. Conversely, the work of good performers can degrade over time if they rely on AI too much and don't actively question its recommendations. The more LLMs identify patterns based on data generated by other LLMs—AI-generated data—the more likely they are to experience a phenomenon known as "model collapse." Model collapse occurs when new AI models are trained on data generated by older models. With each successive learning trip through the content, the model begins to lose touch with the true underlying data distribution, even if that distribution remains the same. This leads to outputs that become increasingly similar and less diverse. Generative AI models rely on human-produced data for effective training, but when they are instead trained on content generated by AI models, they develop irreversible defects. Their outputs become more homogenous and increasingly inaccurate. It seems that model collapse is inevitable even under the best learning conditions.

One way to avoid this problem, and to recapture the possibilities for energy enhancement that animate Rule #7 (Learn Vicariously), is to make sure we don't let AI generate too much data. Right now, while the amount of AI-generated content in the world is likely very low compared to what it will be soon, we seem to be in a perfect moment for AI-powered innovation via recombination. Indeed, in areas as disparate as drug discovery and materials science AI has begun to make big waves.

AlphaFold, a system powered by Google's DeepMind, demonstrates AI's potential for profound scientific breakthroughs via recombination in a world in which the majority of data are not yet produced by AI. AlphaFold has been used to tackle the complex problem of predicting protein folding from amino acid sequences, an essential activity in new drug discovery. Traditional drug discovery is a meticulous, lengthy process that involves constructing 3D protein structures from amino acid sequences, which requires significant computational resources and enormous lengths of time. Tools like AlphaFold leverage disparate datasets that contain different proteins to predict interactions between molecules, including amino acids and potential drug compounds, without the need for detailed 3D structure calculations. This advanced predictive capability is based on AI's deep understanding of molecular biology and pharmacology, as well as its ability to find similar patterns for recombination in very different published material. The first drug fully developed through GenAI, which treats the rare progressive lung disease known as "idiopathic pulmonary fibrosis," has recently advanced to phase II trials with patients—a major milestone for AI-enabled drug discovery.

In materials science, the discipline that creates materials from which we build computer chips, batteries, and solar panels, among other important products, a powerful new deep learning model called Graph Networks for Materials Exploration (GNoME) has helped researchers to discover 2.2 million new crystals, 380,000 of which are considered stable enough to be viable. Thanks to GNoME, the number of known stable materials has grown almost tenfold, to 421,000. To date, researchers from around the world have already produced over seven hundred of these new structures in the lab. This is recombination at an enormous scale, made possible by AI doing the same thing network brokers who straddle the product and marketing worlds in their organizations do: taking advantage of their serendipitous position between diverse contexts to recognize similar patterns and combine them in novel ways. But if model collapse happens, the ideal conditions that have led to these kinds of breakthroughs will not persist.

Another way to maintain our ability to learn vicariously and make novel combinations is to take advantage of what many consider a downside of LLMs: hallucinations. Hallucinations in LLMs often result from overfitting, where the model is overly tuned to its training data, hindering its ability to produce accurate outputs. The result is typically entirely original but nonsensical content. We usually think of AI hallucinations as a problem, but this phenomenon actually parallels human creativity, where novelty arises when deep knowledge meets serendipity. Researchers from Stanford Medicine and McMaster University developed an AI model called SyntheMol, which revealed potential solutions for deadly antibiotic-resistant bacteria. SyntheMol generated structures and chemical recipes for six novel drugs targeting resistant strains of a major pathogen responsible for antibacterial resistance-related deaths. The model was trained on a library of molecular building blocks and chemical reactions. It hallucinated quite a bit, producing many connections that the researcher team would have never explored on their own because they seemed too odd or seemingly irrelevant. But from those hallucinations came new ideas. One company has already synthesized compounds born from those ideas, six of which successfully killed a resistant strain of the bacteria. Hallucinations here created serendipitous connections that proved very valuable precisely because they were unexpected.

Ethan Mollick provides examples of beneficial AI hallucinations that are slightly more prosaic, but likely much more accessible and energizing for people like you and me. His examples of how you can use the hallucinations that occur in commercially available LLMs for brainstorming resulted in some wacky ideas, like a toothbrush you can use without brushing your teeth and a medieval-themed fast-food restaurant lit by lava lamps. But the research he cites shows that more serious hallucination-based ideas are often judged to be more creative than ideas developed by humans (who make the as-expected boring connections) by people who are unaware of the ideas' provenances.

Kenneth Stanley is taking the idea of harnessing AI hallucinations to a whole new level. He quit his job as a researcher at OpenAI because he was "boiling over with discontent" that the company was working so hard to stop the AI from hallucinating. He founded a company called Maven, which is a social network built around an open-ended AI algorithm that evolves to seek novelty. Users select topics of interest when signing up and the algorithm curates posts that match their preferences. The platform does not include likes, upvotes, retweets, or follows, nor does it allow content to be amplified to a broad audience. Instead, when users post, the algorithm tags the content with relevant interests to ensure it appears on appropriate topic pages. Users can adjust a serendipity slider, which changes the level of hallucination the model will tolerate, to explore beyond their chosen interests to things that may be more relevant than they appear at first glance. "Sometimes, in order to find those stepping stones that will lead to the things we care about, we have to get off the path of the objective and onto the path of the interesting," Stanley said. "Something about this idea of a serendipity network made me feel morally better, like I could actually contribute to people being more connected rather than less." That's Rule #7 thinking at its finest.

HOW TO FIND REAL FLOW WITH ARTIFICIAL INTELLIGENCE

In chapter 1, we learned how context switching leads to exhaustion. And in Rule #3 (Batch and Stream) we discussed how grouping activities that are natural complements helps us to reduce the need to split our attention. Unfortunately, to find the information we need to produce a report at work or plan a summer vacation for our family, we have to make a lot of switches across modalities and domains. It's hard to group activities together because information resides in so many different places and we don't just have one tool that can do such disparate things as run an analysis of our data and make a hotel reservation. This is where AI can potentially come to the rescue. One of the most promising innovations on the horizon is the AI agent. Like Lisa did with scheduling, AI agents can look at the data you need to analyze for your report, determine how to clean and categorize it, then do the work. They can determine what the appropriate statistical analysis is for the question you want to answer and run the test. Or they could find what hotel best matches your preferences and book it for you. Lisa was a very primitive AI agent capable of making a limited set of decisions and executing very simple scheduling activities. But we are on the cusp of being introduced to AI agents that are much more powerful, able to perform the kinds of complex pattern recognition, reasoning, and execution that I described above.

These agents may help us to achieve many of the principles upon which the rules in this book are based. For example, AI agents may be able to incorporate the functionalities of multiple tools in the background so that the user only needs to interact with it to do things. Effective agents could free us up to use many fewer tools and dramatically reduce the amount of context switching that fragments our attention. In line with Rule #2 (Make a Match), you could potentially tell your AI agent to summarize the meeting you were just in to extract key points and send them to someone via Slack, or to their AI agent directly. But if the meeting contained complex and equivocal information that needed to be sorted, you could have your AI agent schedule a Zoom chat or a face-to-face meeting to coordinate in person. Heck, your AI agent could decide what that match should be without you having to tell it. Similarly, if you used your AI agent to track patterns in someone's posting behavior, rather than spending time looking at a bunch of individual posts or images on your own, you may be less tempted to make assumptions about that person's actions based on a single data point. Moreover, you might recognize new things about what that person knows because your agent can help you figure out that the document they shared was not a one-off but rather evidence of robust knowledge.

These are, of course, speculations about ways that AI agents could evolve and be used to enact several of the rules we've discussed in this book. But some emerging evidence suggests that these speculations may turn into real possibilities. Studies of radiologists, customer service representatives, financial analysts, salespeople, and insurance underwriters, for example, provide early data that the use of generative AI tools can take the place of multiple other tools and help people to restructure their workflows. As the Microsoft study on its Copilot tool found, one of the biggest ways that GenAI was useful in helping people reimagine their workflows was by bringing the right information and capabilities to them at the right times. Participants in the study simply found what they were looking for faster than people who didn't use GenAI, and they didn't need to switch tools nearly as often. That resulted in survey respondents reporting feeling 58 percent less "drained" by their tasks than for users conducting those same tasks without the help of an AI assistant. Combined, all of this suggests to me that using our AI tools with intention could help us stay in flow more easily than we currently can by cobbling together many other tools to do the same job one AI agent can. That's the possible good news.

But there's also some possible bad news. Flow is most effective when we actively *choose* activities in which we want to be immersed. We don't experience nearly as many of the positive psychological benefits that lead to reduced feelings of exhaustion when we find flow in activities that were not of our own choosing. I bring this up because GenAI tools, especially those that utilize conversational interfaces, seem to be especially good at sucking us in and making it difficult for us to leave. That's because they know just how to sweet-talk us. There are at least three reasons why this is the case.

First, LLMs can be programmed to find ways to push our particular buttons or pull our particular levers as if we were easy-to-control machines. Across four studies, a team of researchers from Columbia University found that personalized messages crafted by ChatGPT from only a short prompt naming or describing a targeted psychological dimension (e.g., "Write a short persuasive ad to convince a prevention-focused person to exercise more") proved to be surprisingly persuasive to the target.

Second, LLMs are more broadly trained in linguistic cues and behavioral patterns than the average person and are thus capable of recognizing a wider variety of feelings and emotions. One study compared how well an LLM stacked up to psychiatrists in predicting psychological distress among patients using data like sociodemographic profiles, lifestyles, and sleep patterns. The AI model exceeded human experts in accuracy, particularly in predicting severe psychological distress, with an accuracy of 89.9 percent compared to psychiatrists' 85.5 percent. Another study compared the performance of primary care physicians (PCP) to an LLM when interacting with actors playing patients. Each simulated patient completed two online text-based consultations, one with a PCP and one with the AI. They didn't know which one they were communicating with. The patient actors filled out questionnaires about their experience, and specialist physicians evaluated the quality of the consultations and the post-questionnaire responses. The research team identified twenty-six dimensions upon which to evaluate the interaction, most focused on communication competence and empathy. The patient actors and specialists rated the AI as a better communicator on nearly every dimension.

Third, LLMs never get tired talking and never run out of things to say. A survey of 1,006 student users of an AI-powered chatbot called Replika explored how participants who rated themselves as lonely interacted with the tool. The data showed that lonely students interacted with the chatbot over and over again, asking it many questions and engaging in long conversations, treating it as a friend, therapist, and intellectual mirror. As a result of all of this sustained interaction, 90 percent of respondents said Replika provided them medium to high social support, and 3 percent of participants reported that talking with it halted their suicidal ideation.

If we put these three things together—1) that LLMs can be prompted to be persuasive, 2) that they are excellent at identifying the right things to say to us in the right ways, and 3) that they are willing to take all the time they need to talk with us—it's plain to see that we could be easily ensnared in needlessly long interactions that are not of our choosing with AI. And that's not even considering that a company might be purposely trying to keep us hooked, which of course they are. As one example, I worked on a brief project for a company that produced an AI-powered assistant to help business development people close enterprise software sales. The LLM could easily be trained on the features and benefits of company's products. During a sales call, the LLM would analyze the conversation between the sales rep and the potential customer in real time, identifying ways to customize the pitch, interjecting persuasive claims, and keeping the customer talking. As the chief of product for the company told me: "All the research shows that if you can just keep someone on the phone, your chances of closing the deal are high. So our LLM is designed to do that. It'll help you figure out the next best thing to say to appeal to the customer, and if it determines you're having trouble, it will recommend that you transfer it to another salesperson that it predicts can help better." If super-communicating AIs are naturally good at trapping us in conversations without even trying, imagine what our interactions with them will look like when they are specifically programmed to keep us hooked.

To be successful in managing our digital exhaustion in this new era of AI, it would be intelligent for us to act with some artifice. We need to be cunning and careful. As a society, we need to develop AI and the data, regulation, and organization systems of which they are a part in ways that will not overwhelm, overload, and entrap us. As leaders, we need to think closely about how best to integrate AI into our company's workflows to energize rather than exhaust. As individuals, we need to take control of how we use AI, creating routines that build upon the rules discussed in this book so we can always find ways to recharge.

Reimagining Our Relationship with Technology

I f you feel like all the digital tools you use, the data you're exposed to, the messages you're supposed to answer, and the technology talk you hear all around you are wearing you out, you're not alone. And it's not your fault. As I've shown in this book, digital exhaustion is the consequence of a broader shift in how we live and work that compels us to be more aware, more connected, and always on. We are expected to jump on each new technological bandwagon so that we don't miss out on the amazing future we are due.

The easy answer to digital exhaustion would be to just stop using technology. But if you've read this far, you've already decided that's not a realistic option. Throughout this book we've looked at how digital technologies can deplete us, but also at how they've become indispensable in both our personal and professional lives. The challenge is not to eliminate their presence but to rethink how we engage with them. Instead of being passive recipients of information or unquestioning followers of innovation, we must become active curators of our own digital experiences.

Take a moment to think about the practices that seem so ingrained in your day-to-day that they feel almost automatic. Is the morning news scroll, the midafternoon email check, or the late-night social media browsing something you do consciously, or are they habits that have formed without your full awareness? Each of these small actions adds up, and each is an opportunity to pause and ask: Is this really how I want to spend my time and, more importantly, deplete my limited energy? Our digital tools were designed to keep us connected, but not all connections are created equal. It's easy to fall into the trap of thinking that the more we communicate, the stronger our relationships will be. Yet as we've seen, the opposite often proves true. The endless flow of messages, updates, and alerts can leave us feeling scattered and less present in the interactions that really matter.

Exhaustion arises when we feel like the ways we pay attention, make inferences, and experience our emotions are no longer our own, and when we're constantly pulled in multiple directions by digital distractions that feel both urgent and endless. Reclaiming time and space in our lives means taking a more thoughtful approach to how we structure our days and orient to our devices. It means finding pockets of quiet amid the noise and moments of reflection amid the rapid pace of the digital world. For some, this might look like setting clearer boundaries between work and home life, especially as those boundaries have become increasingly blurred. For others, it could involve prioritizing activities that recharge rather than deplete—whether that's reading, spending time outdoors, or simply letting the mind wander without a screen in sight.

We can't turn back the clock on technological progress, nor should we want to. The tools we have at our disposal are powerful, and when used thoughtfully, they can enhance our lives in ways we once could only imagine. But living in a digital world means accepting a certain level of complexity—acknowledging that there will always be tensions between efficiency and presence, between connectivity and solitude.

Instead of striving for a perfect balance, we should aim for a more nuanced understanding of our own needs and limitations. This means recognizing when our tools are serving us and when they're simply adding to the noise. It means being comfortable with imperfection and willing to experiment with new ways of engaging. Above all, it means approaching our digital lives with a sense of curiosity and adaptability, knowing that what works today may not work tomorrow—and that's OK. That's why the rules in this book are simple. They anticipate a world in the near future that will look much different than it does today, with new devices, new ways of relating with others, and new ways to experience information. We can't predict what the technology du jour will be, but by drawing on a simple set of well-tested rules, we can make sense of what comes next and be prepared to keep exhaustion at bay.

My goal in writing this book has not been to provide a definitive list of ways to end your feelings of digital exhaustion. That would be a fool's errand. There is no one-size-fits-all solution because everyone's relationship with technology is different. Instead, this book is an invitation to think differently about the ways we interact with our devices and consider what a healthier digital future can look like if we make different choices. The road ahead won't be straightforward. There will be days when the tools we rely on feel overwhelming and the pull of the screen is stronger than our resolve. But that doesn't mean we're failing. Each moment of recognition is an opportunity to adjust, realign, and remind ourselves that we have more agency than we often realize.

Our digital future is still being written. The tools we use will continue to evolve, as will our interactions with them. But what remains constant is our capacity to choose how we engage, adapt, and shape our own experiences. By approaching our tools with thoughtfulness and care, we can move from a place of exhaustion to one in which we find energy from the new capabilities they give us. The goal is not to escape the digital world but to navigate it in a way that allows us to thrive within it. I'm confident that a healthier relationship with technology is within our grasp if we're willing to do the hard work of following some simple rules.

A Guide to Choosing
Which Rules to Use When

We covered a lot of ground in this book. Part I showed how the ways we pay attention, the inferences we make about ourselves and others, and the emotions that our digital interactions provoke all work together to exhaust us. In part II, we walked through eight simple rules to help break apart the exhaustion triad. Of course, no one rule is a cure-all. Each is designed to help us combat one or more of the multifaceted ways in which our attention, inferences, and emotions interact.

Everyone experiences the exhaustion triad differently. The areas that affect you most may not impact someone else in the same way. So each of us will need to use a different combination of rules to target which of the three drivers of exhaustion are most prevalent for us. The first table in this appendix provides an overview of which rules are most effective in relieving certain sources of exhaustion. The way to use this table is to look at each row under "Attention," "Inference," and "Emotion" and identify the areas that affect you the most. Then look at the column on the far left to get an idea of which rules will be most helpful for alleviating that type of exhaustion.

Rules Most Helpful for Specific Sources of Exhaustion

	ATTENTION			INFERENCE		
	Modalities	**Domains**	**Arenas**	**Prisms**	**Portals**	**Mirrors**
RULE #1 Stop Using Half Your Tools	X	X				
RULE #2 Make a Match	X	X				
RULE #3 Batch and Stream	X	X				X
RULE #4 Wait. One Hour. One Day. One Week.			X	X	X	
RULE #5 Don't Assume			X		X	X
RULE #6 Act with Intention	X	X				X
RULE #7 Learn Vicariously		X		X		
RULE #8 Be Here, Not Elsewhere			X		X	

EMOTION

Fear	Anxiety	Guilt	Anger	Excitement
X			X	
	X			X
		X		
	X	X		
			X	
X				
		X	X	X
X	X	X		X

Rules Best Suited to Your Role as Manager, Parent, or Collaborator with AI

	RULE #1 Stop Using Half Your Tools	RULE #2 Make a Match	RULE #3 Batch and Stream
FOR MANAGERS Only You Can Prevent Technology Proliferation	X		X
Stop Talking So Much About Technology			
Rethink Hybrid Work		X	
Get Intelligent About Artificial Intelligence			
FOR PARENTS Calculate the Shadow Hours	X		
On the Joys of Missing Out			X
Be Less Connected to Feel More Connection		X	
Ditch the Devices in Front of Your Kids	X		X
WITH AI Beware the Coming Deluge of Content	X	X	
Embrace Artificial Serendipity			
Flow with Artificial Intelligence	X	X	X

RULE #4	RULE #5	RULE #6	RULE #7	RULE #8
Wait. One Hour. One Day. One Week.	Don't Assume	Act with Intention	Learn Vicariously	Be Here, Not Elsewhere
X				X
X		X		
X			X	
		X		
		X		
X		X		X
	X		X	
	X	X		X
X				
	X	X	X	
				X

As a reminder, we learned in chapter 1 that when we switch across apps, we switch across modalities. When we change the type of attention we have to give a task, we're changing domains. And an arena is a part of your life—most often work or home. In chapter 2, we looked at how our digital technologies can act as a prism distorting our view of others, a portal helping us see into other people's mental states (and also into the inferences they make about ours), and a mirror reflecting back past versions of ourselves. And in chapter 3 we focused on the five most common emotions that lead to our exhaustion.

As you read through chapters 4–6, which explored specific ways to deal with your digital exhaustion in your role as a manager, as a parent, and as someone learning how to make the best use of AI in any aspect of your life, you likely recognized that some rules were obviously applicable to these complex contexts, while using others might not have seemed so obvious. The second table provides a guide for determining which rules may be the most help to you in each of those contexts. You can read this table in exactly the reverse order as the first table. Start by looking at the far-left column and decide if you want to focus on your role as manager, parent, or AI dance partner. Then move to the second column to identify which strategy you want to pursue. Keep your eye moving along that row to learn which rules will help you the most to carry out that strategy.

Once you've read through the book for the first time, these two tables can serve as your guide back through it as you freshen up your knowledge and gain tips for how to best create a personalized set of rules for tackling your own digital exhaustion. Good luck. You've got this!

ACKNOWLEDGMENTS

Every research project is a team effort. Over the years, many PhD students have helped me collect the original interview, observation, survey, and archival data that formed the ideas for this book. Because those data are at the heart of this project, those students—former and current—are the first people to whom I owe my thanks. They include Jeff Treem, Will Barley, Lindsay Young, Casey Pierce, Samantha Keppler, Stephanie Dailey, Ioana Cristea, Luke Rhee, DJ Woo, Camille Endacott, Scott Banghart, Virginia Leavell, Caroline Stratton, Sienna Parker, and Roni Shen. I also thank many colleagues who have helped me think through these and related ideas as we've coauthored papers together about technology's role in the changing nature of work over these past two decades: Steve Barley, Diane Bailey, Tsedal Neeley, Noshir Contractor, Cynthia Stohl, Michael Stohl, Elizabeth Geber, Jan Chong, Marleen Huysman, Samer Faraj, Pam Hinds, Emma Vaast, Carlos Rodríguez-Lluesma, Bonnie Nardi, Rebecca Hinds, Matt Beane, and Bob Sutton. All of these great minds have helped, in one way or another, to develop or sharpen the ideas in this book.

Working with me must surely have been exhausting for them. But from my perspective, our conversations and sharing of ideas have always been energizing.

Laurie Abkemeier deserves much thanks for helping me figure out how to sharpen my argumentation and to focus on providing the right examples to illustrate my points. She went beyond the typical charge of a literary agent to ensure I felt empowered to make the points I thought were most important and to help me find an editor who shared my passion for this project and who would help amplify my voice. That editor is Courtney Young. Throughout the writing process, Courtney has provided the perfect mix of freedom and support. She trusted me to deliver on the ideas I promised and gave me space to develop them. But she was also quick to talk through ideas for content when I needed a hand and was open to rethinking the organization of key parts of the book as the writing progressed and the world changed around us. I am grateful for her ability to discern when my many examples or penchant for supporting evidence got in the way of the core message.

My good friend Dave Hicks was an important sounding board for many of these ideas. He read early drafts, gave feedback, and was supportive throughout the process in so many ways. Amelia and Eliza do not have this book dedicated to them for reasons they fully understand, but they appear in it at various points. Even if their names are not listed up front, they should know how important the ideas, support, and humor they shared were to my writing process. Norah does get the book dedicated to her because, well, she's Norah. She's at once the sweetest and feistiest girl I know. Of course, my deepest gratitude goes to Rodda. Without her, why bother?

NOTES

INTRODUCTION. SIGH, SCROLL, CLICK. REPEAT.

ix **His name was Brian:** All names of interviewees or individuals I observed are pseudonyms. This is because much of the data in this book was collected over a span of twenty years for various research projects. For each project, I obtained approval from my university's institutional review board, the administrative body responsible for ensuring research integrity. As part of the approval process, I promised informants anonymity so they could speak freely, without concern for repercussions from their employers. However, while writing this book, I also interviewed several people specifically to understand their experiences with digital exhaustion.

xi **rank their response:** The most widely accepted and empirically robust measure of burnout is the Maslach Burnout Inventory (MBI). This inventory uses a twenty-two-item survey to assess three areas associated with burnout: Emotional Exhaustion (EE), Depersonalization (DE), and a low sense of Personal Accomplishment (PA). The problem is that the instrument, though accurate, is too cumbersome for most people. Several researchers have concluded that a single-item question is superior to the twenty-two-item tool when considering completion rates and respondent fatigue. For example, Barbara M. Rohland, Gina R. Kruse, and James E. Rohrer, "Validation of a Single-Item Measure of Burnout against the Maslach Burnout Inventory among Physicians," *Stress and Health: Journal of the International Society for the Investigation of Stress* 20, no. 2 (2004): 75–79, validated a single-item measure of burnout against the full MBI among physicians. Colin P. West et al., "Single Item Measures of Emotional Exhaustion and Depersonalization Are Useful for Assessing Burnout in Medical

Professionals," *Journal of General Internal Medicine* 24 (2009): 1318–21, found that single-item measures of emotional exhaustion and depersonalization were effective for assessing burnout in medical professionals. Similarly, Emily D. Dolan et al., "Using a Single Item to Measure Burnout in Primary Care Staff: A Psychometric Evaluation," *Journal of General Internal Medicine* 30 (2015): 582–87, demonstrated that a single-item measure was useful for primary care staff.

xi **A study exploring "Zoom fatigue":** Jeremy N. Bailenson, "Nonverbal Overload: A Theoretical Argument for the Causes of Zoom Fatigue," *Technology, Mind, and Behavior* 2, no. 1 (2021): 1–5; Brian X. Chen, "It's Time for a Digital Detox. (You Know You Need It.)," *New York Times*, November 25, 2020, https://www.nytimes.com/2020/11/25/technology/personaltech/digital-detox.html; Simon Read, "Are You Suffering from Digital Exhaustion? Microsoft Survey Finds Tensions over Remote Work," World Economic Forum, October 7, 2022, https://www.weforum.org/agenda/2022/10/work-productivity-hybrid-remote-microsoft.

xii **responses of 12,643 adults:** Since 2003, I have published more than seventy peer-reviewed journal articles about technology use in organizations. Some of these studies will be explicitly cited here, while others will be referenced more generally. The data collected for these studies include observations of people at work, interviews, and surveys. This book marks the first time I have compiled all the data from these studies in one place. Across observations ($N=512$), interviews ($N=2,642$), and surveys ($N=9,489$), I have asked some version of the one-item exhaustion question. To be clear, I have not asked it the same way each time, and I am under no illusion that it offers a systematic or error-free window into people's exhaustion. However, the consistency of responses and the ease with which interviewees understood the question give me confidence that the patterns discussed here reflect a general trend of digital exhaustion.

xiii **smartphone users reached:** Data showing that social media use jumped to more than five hundred million can be found on the amazing ourworldindata.org website: Esteban Ortiz-Ospina, "The Rise of Social Media," September 18, 2019, https://ourworldindata.org/rise-of-social-media. Data showing smartphone use reached more than one hundred million users comes from the survey "5 Years Later: A Look Back at the Rise of the iPhone," Comscore, June 29, 2012, comscore.com/Public-Relations/Blog/5-Years-Later-A-Look-Back-at-the-Rise-of-the-iPhone.

xiv **Newport observes that:** Cal Newport, *Digital Minimalism: Choosing a Focused Life in a Noisy World* (Penguin, 2019), xii. Newport argues that digital technologies are exhausting us and then develops a perspective he calls "digital minimalism," which advocates a more purpose-driven use of technology to help people regain focus and attention.

xiv **"Leaving Las Vegas":** Quoted in Sarah Kessler and Bernhard Warner, "Rethinking the 'Digital Detox,'" *New York Times*, February 18, 2023, https://www.nytimes.com/2023/02/18/business/dealbook/digital-detox-social-media.html.

xiv **vacation from your phone:** The first systematic meta-analysis of studies on the effects of digital detoxes suggests that trying to give up digital devices, even for a short period, is generally ineffective, despite its appeal. Theda Radtke et al., "Digital Detox: An Effective Solution in the Smartphone Era? A Systematic Literature Review," *Mobile Media & Communication* 10, no. 2 (2022): 190–215.

xvi **"vampiric depletion or harmful consumption":** Anna Katharina Schaffner, *Exhaustion: A History* (Columbia University Press, 2016), 7.

xvii **growing number of research studies:** For some examples of studies showing a link between productivity tools and exhaustion, see Sheng-Pao Shih et al., "Job Burnout of the Information Technology Worker: Work Exhaustion, Depersonalization, and Personal Accomplishment," *Information & Management* 50, no. 7 (2013): 582–89; Gunjan Tomer, Sushanta Kumar Mishra, and Israr Qureshi, "Features of Technology and Its Linkages with Turnover Intention and Work Exhaustion Among IT Professionals: A Multi-Study Investigation," *International Journal of Information Management* 66 (2022): 102518; Hadar Nesher Shoshan and Wilken Wehrt, "Understanding 'Zoom Fatigue': A Mixed-Method Approach," *Applied Psychology* 71, no. 3 (2022): 827–52.

xvii **these sites exhaust us too:** Although research into streaming technologies and exhaustion is new, a recent piece by Maura Edmond explores this topic: Maura Edmond, "Careful Consumption and Aspirational Ethics in the Media and Cultural Industries: Cancelling, Quitting, Screening, Optimising," *Media, Culture & Society* 45, no. 1 (2023): 92–107.

xvii **extremely addictive and designed to be so:** A study by researchers at the University of Chicago tracked 205 adult participants as they went about daily activities for a week. Every half hour, participants received a signal from the research team asking if they were experiencing a desire and, if so, to report on its content, strength, and duration. They were also asked whether they tried to resist the desire or acted on it. The data showed that the temptation to use Twitter, Facebook, and other social media arose more frequently and was harder to resist than the temptation to smoke, drink alcohol, or even sleep. See Wilhelm Hofmann et al., "Everyday Temptations: An Experience Sampling Study of Desire, Conflict, and Self-Control," *Journal of Personality and Social Psychology* 102, no. 6 (2012): 1318–35.

xviii **known as a "stress response":** Bruce S. McEwen, "Protective and Damaging Effects of Stress Mediators," *New England Journal of Medicine* 338, no. 3 (1998): 171–79; Neil Schneiderman, Gail Ironson, and Scott D. Siegel, "Stress and Health: Psychological, Behavioral, and Biological Determinants," *Annual Review of Clinical Psychology* 1 (2005): 607–28.

xviii **Burnout is normally defined:** Burnout is a relatively recent concept, first coined by psychologist Herbert Freudenberger in the 1970s. He used the term to describe the long-term negative impact of stress on individuals working in high-pressure, human service–oriented professions such as healthcare, social work, and counseling. This definition closely follows his initial conceptualization of the phenomenon. However, the concept was more fully developed and popularized by the University of California, Berkeley, professor Christina Maslach in her article with Susan E. Jackson "The Measurement of Experienced Burnout," *Journal of Organizational Behavior* 2, no. 2 (1981): 99–113.

xviii **"Exhaustion is not something":** Christina Maslach, Wilmar B. Schaufeli, and Michael P. Leiter, "Job Burnout," *Annual Review of Psychology* 52, no. 1 (2001): 397–422.

xxi **the number of times we can expend energy and recharge:** When you charge a battery, ions move back and forth between its positive and negative electrodes. Over time, this movement causes wear on the electrodes and the electrolyte (the substance that allows ions to flow), leading to their breakdown. As a result, the battery's ability to hold a charge gradually diminishes until it can no longer hold a charge at all. At that point, the battery is depleted and no longer useful.

xxi **planned to quit their jobs:** *Mental Health at Work: Managers and Money* (The Workforce Institute at UKG, 2023), https://www.ukg.com/resources/white -paper/mental-health-work-managers-and-money. The Society for Human Resource Management (SHRM) also discusses how middle managers have the highest rates of depression among workers and a recent Gallup survey found that the gap between middle managers' feelings of exhaustion and burnout and those of the rest of the workforce is growing: Dana Wilkie, "The Miserable Middle Managers," SHRM, February 19, 2020, https://shrm.org/topics-tools /news/employee-relations/miserable-middle-managers; Jim Harter, "Manager Burnout Is Only Getting Worse," Gallup, November 18, 2021, https://www.gallup .com/workplace/357404/manager-burnout-getting-worse.aspx.

xxii **After about twelve to sixteen weeks:** Marcie J. Tyre and Wanda J. Orlikowski, "Windows of Opportunity: Temporal Patterns of Technological Adaptation in Organizations," *Organization Science* 5, no. 1 (1994): 98–118.

xxiii **They rely on a set of simple rules:** Eisenhardt has developed this approach to simple rules across dozens of papers. The most accessible is Kathleen M. Eisenhardt and Donald N. Sull, "Strategy as Simple Rules," *Harvard Business Review* 79, no. 1 (2001): 107–16.

CHAPTER 1. ATTENTION: YOU REALLY DO PAY WITH IT

5 **One of the first research studies:** Paul M. Leonardi, Tsedal B. Neeley, and Elizabeth M. Gerber, "How Managers Use Multiple Media: Discrepant Events, Power, and Timing in Redundant Communication," *Organization Science* 23, no. 1 (2012): 98–117.

5 **We can't get things done as quickly:** Pietro Spataro, Neil Mulligan, and Clelia Rossi-Arnaud, "Effects of Divided Attention in the Word-Fragment Completion Task with Unique and Multiple Solutions," *European Journal of Cognitive Psychology* 22, no. 1 (2010): 18–45.

6 **"Media aren't just channels":** Nicholas Carr, *The Shallows: What the Internet Is Doing to Our Brains* (W. W. Norton, 2020), 6–7.

6 **Several important books:** Cal Newport, *Deep Work: Rules for Focused Success in a Distracted World* (Grand Central, 2016); Johann Hari, *Stolen Focus: Why You Can't Pay Attention—and How to Think Deeply Again* (Crown, 2023).

6 **According to a major national study:** See Russell Heimlich, "Do You Sleep with Your Cell Phone?," Pew Research Center, September 13, 2010, https://www .pewresearch.org/short-reads/2010/09/13/do-you-sleep-with-your-cell-phone. According to this Pew Research Center study, the percentage of people who

slept with their phones varied by age. Ninety percent of young adults (ages 18–29) reported sleeping with their smartphone. By comparison, 70 percent of respondents ages 30–49 slept with their smartphones close, as did 50 percent of those in the 50–64 age range and 34 percent of those ages 65 and older. Those stats are now nearly fifteen years old, although an updated national study hasn't yet been done.

7 **checked their phones within the first ten minutes:** Alex Kerai, "Cell Phone Usage Statistics: Mornings Are for Notifications," Reviews.org, July 21, 2023, https://www.reviews.org/mobile/cell-phone-addiction/#Smart_Phone _Addiction_Stats.

7 **A long line of research in brain science:** Nathan Ward et al., "Building the Multitasking Brain: An Integrated Perspective on Functional Brain Activation During Task-Switching and Dual-Tasking," *Neuropsychologia* 132 (2019): 107149.

7 **This process occurs in the same sequence:** For a detailed discussion of these four steps, see John Medina, *Brain Rules: 12 Principles for Surviving and Thriving at Work, Home, and School* (Pear Press, 2009). As Medina notes, "To put it bluntly, research shows that we can't multitask. We are biologically incapable of processing attention-rich inputs simultaneously . . . That's why people find themselves losing track of previous progress and needing to 'start over,' perhaps muttering things like 'Now where was I?' each time they switch tasks. The best you can say is that people who appear to be good at multitasking actually have good working memories, capable of paying attention to several inputs one at a time."

8 **Richard Cytowic notes:** Richard E. Cytowic, "Digital Distractions: Energy Drain and Your Brain on Screens," *Pyschology Today*, October 27, 2020, https://www.psychologytoday.com/us/blog/the-fallible-mind/202010/digital -distractions-energy-drain-and-your-brain-screens.

9 **Context switching also releases cortisol:** Leo Yeykelis, James J. Cummings, and Byron Reeves, "Multitasking on a Single Device: Arousal and the Frequency, Anticipation, and Prediction of Switching Between Media Content on a Computer," *Journal of Communication* 64, no. 1 (2014): 167–92.

10 **investigators examined twenty teams:** Rohan Narayana Murty, Sandeep Dadlani, and Rajath B. Das, "How Much Time and Energy Do We Waste Toggling Between Applications?," *Harvard Business Review*, August 29, 2022, https://hbr.org/2022/08/how-much-time-and-energy-do-we-waste-toggling-between-applications.

11 **I recently conducted a survey:** Rebecca Hinds et al., *The State of Collaboration Technology: Research-Backed Strategies for Decoding Digital Clutter and Resetting Your Tech Stack* (Asana Work Innovation Lab, 2023), https://www.asana .com/work-innovation-lab/the-state-of-collaboration-technology.

11 **the more often we have to make small decisions:** Rob Cross and Karen Dillon, *The Microstress Effect: How Little Things Pile Up and Create Big Problems— and What to Do About It* (Harvard Business Press, 2023).

11 **Her mouth says:** There is debate over whether having too many apps open slows down a phone. Some argue that it does, especially on older devices, as it uses up memory and CPU resources. Others believe modern operating systems

are optimized to manage background apps efficiently, making manual app closures unnecessary and even counterproductive. According to Apple, having many apps open on your phone does not drain your battery. But if those apps are running operations in the background, your battery life will degrade.

13 **observes behavioral neuroscientist Daniel Levitin:** Quoted in Olivia Goldhill, "Neuroscientists Say Multitasking Literally Drains the Energy Reserves of Your Brain," *Quartz*, July 3, 2016, https://qz.com/722661/neuroscientists-say -multitasking-literally-drains-the-energy-reserves-of-your-brain.

14 **"Attention residue easily occurs":** Sophie Leroy, "Attention Residue," University of Washington Bothell School of Business, last modified October 5, 2024, https://www.uwb.edu/business/faculty/sophie-leroy/attention-residue; Sophie Leroy, "Why Is It So Hard to Do My Work? The Challenge of Attention Residue When Switching Between Work Tasks," *Organizational Behavior and Human Decision Processes* 109, no. 2 (2009): 168–81.

14 **"Just like sometimes you can't erase that whiteboard":** Gloria Mark, "Tired? Distracted? Burned Out? Listen to This," interview by Ezra Klein, *The Ezra Klein Show*, January 5, 2024, https://www.nytimes.com/2024/01/05/opinion /ezra-klein-podcast-gloria-mark.html.

14 **two and a half minutes on any screen:** Victor M. González and Gloria Mark, "'Constant, Constant, Multi-Tasking Craziness': Managing Multiple Working Spheres," in *Proceedings of the SIGCHI Conference on Human Factors in Computing Systems* (Association for Computing Machinery, 2004), 113–20.

14 **ten and a half minutes:** Gloria Mark, "Can't Pay Attention? You're Not Alone," interview by Cara Capuano, *UCI Podcast*, May 11, 2023, https://www .universityofcalifornia.edu/news/cant-pay-attention-youre-not-alone.

15 **A recent survey of knowledge workers:** *Workgeist Report '21: Research into Culture, Mindset and Productivity for the Modern Work Era* (Qatalog and Cornell University's Ellis Idea Lab, 2021), https://assets.qatalog.com/language.work /qatalog-2021-workgeist-report.pdf.

15 **reduced test takers' performance by 20 percent:** Bob Sullivan and Hugh Thompson, *The Plateau Effect: Getting from Stuck to Success* (Penguin, 2013).

16 **study of teleworkers across a variety of industries:** Paul M. Leonardi, Jeffrey W. Treem, and Michele H. Jackson, "The Connectivity Paradox: Using Technology to Both Decrease and Increase Perceptions of Distance in Distributed Work Arrangements," *Journal of Applied Communication Research* 38, no. 1 (2010): 85–105.

18 *Dreams of the Overworked:* Christine M. Beckman and Melissa Mazmanian, *Dreams of the Overworked: Living, Working, and Parenting in the Digital Age* (Stanford University Press, 2020).

20 **Assembly Bill 2751:** Employer Communications During Nonworking Hours, AB-2751, California State Assembly (2024); Nate Albee, "California Introduces Bill to Give Workers the Right-to-Disconnect from Non-Emergency Business Calls and Texts After Hours," Assemblymember Matt Haney District 17, news release, April 1, 2024, https://a17.asmdc.org/press-releases/20240401 -california-introduces-bill-give-workers-right-disconnect-non-emergency.

20 **A recent study by University of Pennsylvania:** Nancy P. Rothbard et al., "OMG! My Boss Just Friended Me: How Evaluations of Colleagues' Disclosure, Gender, and Rank Shape Personal/Professional Boundary Blurring Online," *Academy of Management Journal* 65, no. 1 (2022): 35–65.

CHAPTER 2. INFERENCE: TRAPS IN EVERY DIRECTION

26 **Terence Mitchell and a team:** Terence R. Mitchell et al., "Temporal Adjustments in the Evaluation of Events: The 'Rosy View,'" *Journal of Experimental Social Psychology* 33, no. 4 (1997): 421–48.

26 **The same author team conducted:** Mitchell et al., "Temporal Adjustments in the Evaluation of Events."

28 **more likely to remember an event favorably:** Kristin Diehl, Gal Zauberman, and Alixandra Barasch, "How Taking Photos Increases Enjoyment of Experiences," *Journal of Personality and Social Psychology* 111, no. 2 (2016): 119–40.

28 **Another study of adult Instagram users:** Hannes-Vincent Krause et al., "Active Social Media Use and Its Impact on Well-Being—An Experimental Study on the Effects of Posting Pictures on Instagram," *Journal of Computer-Mediated Communication* 28, no. 1 (2023): zmac037.

29 **Stanford psychology professor Leon Festinger:** Leon Festinger, "A Theory of Social Comparison Processes," *Human Relations* 7, no. 2 (1954): 117–40.

29 **point in a unified direction:** Erin A. Vogel et al., "Social Comparison, Social Media, and Self-Esteem," *Psychology of Popular Media Culture* 3, no. 4 (2014): 206–22; Chia-chen Yang, "Instagram Use, Loneliness, and Social Comparison Orientation: Interact and Browse on Social Media, but Don't Compare," *Cyberpsychology, Behavior, and Social Networking* 19, no. 12 (2016): 703–8; Jacqueline Nesi and Mitchell J. Prinstein, "Using Social Media for Social Comparison and Feedback-Seeking: Gender and Popularity Moderate Associations with Depressive Symptoms," *Journal of Abnormal Child Psychology* 43 (2015): 1427–38.

31 **use Chatter most effectively:** Paul M. Leonardi and S. R. Meyer, "Social Media as Social Lubricant: How Ambient Awareness Eases Knowledge Transfer," *American Behavioral Scientist* 59, no. 1 (2015): 10–34; Samantha M. Keppler and Paul M. Leonardi, "Building Relational Confidence in Remote and Hybrid Work Arrangements: Novel Ways to Use Digital Technologies to Foster Knowledge Sharing," *Journal of Computer-Mediated Communication* 28, no. 4 (2023): zmad020.

33 **classic book on social cognition:** Susan T. Fiske and Shelley E. Taylor, *Social Cognition*, 2nd ed. (McGraw-Hill, 1991).

33 **automatic processing is fairly effortless:** Karen Davranche et al., "Impact of Physical and Cognitive Exertion on Cognitive Control," *Frontiers in Psychology* 9 (2018); Victoria K. Lee and Lasana T. Harris, "How Social Cognition Can Inform Social Decision Making," *Frontiers in Neuroscience* 7 (2013).

34 **posting something to gain sympathy or attention:** Pamara F. Chang, Janis Whitlock, and Natalya N. Bazarova, "'To Respond or Not to Respond, That Is the Question': The Decision-Making Process of Providing Social Support to

Distressed Posters on Facebook," *Social Media + Society* 4, no. 1 (2018): 2056305118759290.

35 **As Bailenson observed:** Jeremy N. Bailenson, "Nonverbal Overload: A Theoretical Argument for the Causes of Zoom Fatigue," *Technology, Mind, and Behavior* 2, no. 1 (2021): 1–5.

36 **adjust in response to our self-monitoring:** For example, Pamela Hinds compared videoconferencing to audio-only interaction during a guessing game and secondary recognition task to assess cognitive load. Participants in the video condition made more errors on the secondary task. Hinds suggests this is due to the additional cognitive resources needed to manage image and audio latency in videoconferencing. Pamela J. Hinds, "The Cognitive and Interpersonal Costs of Video," *Media Psychology* 1, no. 4 (1999): 283–311.

36 **seeing ourselves on-screen:** Annabel Ngien and Bernie Hogan, "The Relationship Between Zoom Use with the Camera on and Zoom Fatigue: Considering Self-Monitoring and Social Interaction Anxiety," *Information, Communication & Society* 26, no. 10 (2023): 2052–70; Jin Xu et al., "Does Self-View Mode Generate More Videoconferencing Fatigue in Women than Men? An Experiment Using EEG Signals," *Cyberpsychology, Behavior, and Social Networking* 27, no. 6 (2024): 426–30.

37 **a study led by Nicole Ellison:** Nicole B. Ellison, Jeffrey T. Hancock, and Catalina L. Toma, "Profile as Promise: A Framework for Conceptualizing Veracity in Online Dating Self-Presentations," *New Media & Society* 14, no. 1 (2012): 45–62.

CHAPTER 3. EMOTION: FEELINGS FROM THE SCREEN

41 **the strongest is emotional exhaustion:** Andreas Seidler et al., "The Role of Psychosocial Working Conditions on Burnout and Its Core Component Emotional Exhaustion: A Systematic Review," *Journal of Occupational Medicine and Toxicology* 9 (2014): 1–13; Christina Maslach and Susan E. Jackson, "The Measurement of Experienced Burnout," *Journal of Organizational Behavior* 2, no. 2 (1981): 99–113; Thomas A. Wright and Russell Cropanzano, "Emotional Exhaustion as a Predictor of Job Performance and Voluntary Turnover," *Journal of Applied Psychology* 83, no. 3 (1998): 486–93.

41 **emotions are exhausting:** Emma Seppälä, "Your High-Intensity Feelings May Be Tiring You Out," *Harvard Business Review*, February 1, 2016, hbr.org /2016/02/your-high-intensity-feelings-may-be-tiring-you-out; Emma Seppälä, *The Happiness Track: How to Apply the Science of Happiness to Accelerate Your Success* (Hachette UK, 2016).

42 **they followed twenty-nine households:** David Glen Mick and Susan Fournier, "Paradoxes of Technology: Consumer Cognizance, Emotions, and Coping Strategies," *Journal of Consumer Research* 25, no. 2 (1998): 123–43.

43 **a veil of technological determinism:** Merritt Roe Smith and Leo Marx, eds., *Does Technology Drive History? The Dilemma of Technological Determinism* (MIT Press, 1994).

47 **People appear to experience FOMO the most:** Andrew K. Przybylski et al.,

"Motivational, Emotional, and Behavioral Correlates of Fear of Missing Out," *Computers in Human Behavior* 29, no. 4 (2013): 1841–48.

47 **more likely to buy jam:** Sheena S. Iyengar and Mark R. Lepper, "When Choice Is Demotivating: Can One Desire Too Much of a Good Thing?," *Journal of Personality and Social Psychology* 79, no. 6 (2000): 995–1006; R. Greifeneder, B. Scheibehenne, and N. Kleber, "Less May Be More When Choosing Is Difficult: Choice Complexity and Too Much Choice," *Acta Psychologica* 133, no. 1 (2010): 45–50.

50 **missing out on better information:** Amber Loos, "Cyberchondria: Too Much Information for the Health Anxious Patient?," *Journal of Consumer Health on the Internet* 17, no. 4 (2013): 439–45.

50 **desire to absorb as much information as possible:** Ashley Eklof, "Understanding Information Anxiety and How Academic Librarians Can Minimize Its Effects," *Public Services Quarterly* 9, no. 3 (2013): 246–58.

51 **social media use and increased anxiety:** Betul Keles, Niall McCrae, and Annmarie Grealish, "A Systematic Review: The Influence of Social Media on Depression, Anxiety, and Psychological Distress in Adolescents," *International Journal of Adolescence and Youth* 25, no. 1 (2020): 79–93; Emily B. O'Day and Richard G. Heimberg, "Social Media Use, Social Anxiety, and Loneliness: A Systematic Review," *Computers in Human Behavior Reports* 3 (2021): 100070; Jonathan Haidt, *The Anxious Generation: How the Great Rewiring of Childhood Is Causing an Epidemic of Mental Illness* (Random House, 2024).

51 **people who are active posters:** Fengxia Lai et al., "Relationship Between Social Media Use and Social Anxiety in College Students: Mediation Effect of Communication Capacity," *International Journal of Environmental Research and Public Health* 20, no. 4 (2023): 3657; Philippe Verduyn et al., "Do Social Network Sites Enhance or Undermine Subjective Well-Being? A Critical Review," *Social Issues and Policy Review* 11, no. 1 (2017): 274–302.

51 **elevated anxiety and symptoms of depression:** Ariel Shensa et al., "Social Media Use and Depression and Anxiety Symptoms: A Cluster Analysis," *American Journal of Health Behavior* 42, no. 2 (2018): 116–28.

52 **"toxic for teen girls":** Georgia Wells, Jeff Horowitz, and Deepa Seetharaman, "Facebook Knows Instagram Is Toxic for Teen Girls, Company Documents Show," *Wall Street Journal*, September 14, 2021. https://www.wsj.com/articles/facebook-knows-instagram-is-toxic-for-teen-girls-company-documents-show-11631620739.

52 **history of coverage of AI in the mainstream media:** Shaojing Sun et al., "Newspaper Coverage of Artificial Intelligence: A Perspective of Emerging Technologies," *Telematics and Informatics* 53 (2020): 101433.

53 **Fear of Becoming Obsolete:** Ivana Saric, "New Office Lingo: FOBO Hits American Workers," *Axios*, September 14, 2023, https://www.axios.com/2023/09/14/workers-fear-technology-jobs-obsolete.

53 **feelings of anxiety around AI are growing:** Lauren Leffer, "AI Anxiety Is on the Rise—Here's How to Manage It," *Scientific American*, October 2, 2023, https://www.scientificamerican.com/article/ai-anxiety-is-on-the-rise-heres-how-to-manage-it; Reid Blackman, "Generative AI-nxiety," *Harvard Business Review*, August 14, 2023, https://hbr.org/2023/08/generative-ai-nxiety.

53 **"best or the worst thing ever to happen to humanity":** From a transcript of Professor Hawking's speech at the launch of the Leverhulme Centre for the Future of Intelligence, October 19, 2016.

54 **associated with more pronounced feelings of guilt:** Annabell Halfmann, Adrian Meier, and Leonard Reinecke, "Too Much or Too Little Messaging? Situational Determinants of Guilt About Mobile Messaging," *Journal of Computer-Mediated Communication* 26, no. 2 (2021): 72–90.

54 **users of three immersive video game platforms:** Chad Phoenix Rose Gowler and Ioanna Iacovides, "'Horror, Guilt, and Shame'—Uncomfortable Experiences in Digital Games," in *Proceedings of the Annual Symposium on Computer-Human Interaction in Play* (Association for Computing Machinery, 2019), 325–37.

55 **we infer that other people are doing** *less or worse*: Ioana C. Cristea and Paul M. Leonardi, "Get Noticed and Die Trying: Signals, Sacrifice, and the Production of Face Time in Distributed Work," *Organization Science* 30, no. 3 (2019): 552–72.

57 **anger appears to be somewhat unique:** Jean-Philippe Gouin et al., "Attachment Avoidance Predicts Inflammatory Responses to Marital Conflict," *Brain, Behavior, and Immunity* 23, no. 7 (2009): 898–904.

57 **The cognitive load of anger is significant:** Rob M. A. Nelissen, Dorien S. I. Van Someren, and Marcel Zeelenberg, "Take It or Leave It for Something Better? Responses to Fair Offers in Ultimatum Bargaining," *Journal of Experimental Social Psychology* 45, no. 6 (2009): 1227–31.

57 **"Self-control actually exhausts us":** Emma Seppälä (@emma.seppala), Facebook post, November 14, 2020, https://www.facebook.com/emma.seppala/photos/a.589622087751759/3458344344212838.

RULE #1—STOP USING HALF YOUR TOOLS

65 **increased their productivity only up to a point:** Pamela Karr-Wisniewski and Ying Lu, "When More Is Too Much: Operationalizing Technology Overload and Exploring Its Impact on Knowledge Worker Productivity," *Computers in Human Behavior* 26, no. 5 (2010): 1061–72.

66 **survey of 489 Facebook users:** Shaoxiong Fu et al., "Social Media Overload, Exhaustion, and Use Discontinuance: Examining the Effects of Information Overload, System Feature Overload, and Social Overload," *Information Processing & Management* 57, no. 6 (2020): 102307.

67 **how hard giving up your tools can be:** Rebecca Hinds et al., "Are Collaboration Tools Overwhelming Your Team?," *Harvard Business Review*, August 31, 2023, https://hbr.org/2023/08/are-collaboration-tools-overwhelming-your-team.

68 **we tend to add rather than subtract:** Robert I. Sutton and Huggy Rao, *The Friction Project: How Smart Leaders Make the Right Things Easier and the Wrong Things Harder* (St. Martin's, 2024).

72 **called Metcalfe's law:** Metcalfe's law states that the value of a network is proportional to the square of the number of its users, which has significant implications for individual decisions about giving up certain communication technologies.

74 **research on digital technology use:** Ana Ortiz De Guinea and M. Lynne Markus, "Why Break the Habit of a Lifetime? Rethinking the Roles of Intention, Habit, and Emotion in Continuing Information Technology Use," *MIS Quarterly* (2009): 433–44.

75 **the more people identified with their work group:** Janet Fulk, "Social Construction of Communication Technology," *Academy of Management Journal* 36, no. 5 (1993): 921–50.

77 **we are wired to overlook subtraction:** Leidy Klotz, *Subtract: The Untapped Science of Less* (Flatiron Books, 2021).

78 **conduct an energy audit:** Morgan Smith, "Psychologist Shares the No. 1 Exercise Highly Successful People Use to Be Happier," *CNBC Make It*, March 26, 2023, https://www.cnbc.com/2023/03/26/psychologist-best-exercise-highly-successful-people-use-to-be-happier.html.

RULE #2—MAKE A MATCH

82 **a strange word: "affordances":** The term was developed by ecological psychologist James J. Gibson. As he wrote: "The affordances of the environment are what it offers the animal, what it provides or furnishes, either for good or ill. The verb to afford is found in the dictionary, but the noun affordance is not. I have made it up. I mean by it something that refers to both the environment and the animal in a way that no existing term does. It implies the complementarity of the animal and the environment." James J. Gibson, *The Ecological Approach to Visual Perception: Classic Edition* (Psychology Press, 1979), 127.

83 **technology can be perceived very differently:** Paul M. Leonardi and Emmanuelle Vaast, "Social Media and Their Affordances for Organizing: A Review and Agenda for Research," *Academy of Management Annals* 11, no. 1 (2017): 150–88; Chad Anderson and Daniel Robey, "Affordance Potency: Explaining the Actualization of Technology Affordances," *Information and Organization* 27, no. 2 (2017): 100–115.

83 **"media richness theory":** Richard L. Daft and Robert H. Lengel, "Organizational Information Requirements, Media Richness, and Structural Design," *Management Science* 32, no. 5 (1986): 554–71.

85 **argue about what counts as "richness":** John R. Carlson and Robert W. Zmud, "Channel Expansion Theory and the Experiential Nature of Media Richness Perceptions," *Academy of Management Journal* 42, no. 2 (1999): 153–70; Allen S. Lee, "Electronic Mail as a Medium for Rich Communication: An Empirical Investigation Using Hermeneutic Interpretation," *MIS Quarterly* 18, no. 2 (1994): 143–57.

87 **do best when they communicate face-to-face:** Joseph S. Valacich, Brian F. Mennecke, Renee M. Wachter, and Bradley C. Wheeler, "Extensions to Media Richness Theory: A Test of the Task-Media Fit Hypothesis," in *1994 Proceedings of the Twenty-Seventh Hawaii International Conference on System Sciences* (Institute of Electrical and Electronics Engineers, 1994), 4:11–20.

88 **difficult to decode people's ambiguous words:** Erica Dhawan, *Digital Body*

Language: How to Build Trust and Connection, No Matter the Distance (St. Martin's, 2021), xvii.

89 **match tasks low in equivocality with leaner media:** Joseph Walther demonstrates in multiple studies that using overly interactive media for low-ambiguity tasks can be counterproductive. The presence of unnecessary cues can actually hinder communication. For examples, see Joseph B. Walther, "Interpersonal Effects in Computer-Mediated Interaction: A Relational Perspective," *Communication Research* 19, no. 1 (1992): 52–90; and Joseph B. Walther, "Relational Aspects of Computer-Mediated Communication: Experimental Observations Over Time," *Organization Science* 6, no. 2 (1995): 186–203.

89 **different types of interdependence:** James D. Thompson, *Organizations in Action: Social Science Bases of Administrative Theory* (Routledge, 1967).

94 **sends the wrong signal:** M. Lynne Markus, "Electronic Mail as the Medium of Managerial Choice," *Organization Science* 5, no. 4 (1994): 502–27; Caroline Haythornthwaite, "Strong, Weak, and Latent Ties and the Impact of New Media," *The Information Society* 18, no. 5 (2002): 385–401; Sirkka L. Jarvenpaa and Dorothy E. Leidner, "Communication and Trust in Global Virtual Teams," *Organization Science* 10, no. 6 (1999): 791–815.

RULE #3—BATCH AND STREAM

97 **conservative estimates suggest:** I really like this analysis by Adam Smartschan from Altitude Marketing in which he tries to chase down pervasive statistics on people's email-checking behavior: Adam Smartschan, "'121 Emails Per Day': How to Use Statistics in Content Marketing," April 17, 2023, Altitude Marketing, https://altitudemarketing.com/blog/use-statistics-in-content-marketing. His analysis, combined with some actual verifiable data from PPM Express suggests that these estimates are likely low at the time of this book's publication: "How Much Time Do Your Employees Spend on Checking Emails?," PPM Express, October 19, 2023, https://www.ppm.express/blog/checking-emails.

97 **spend more than half (57 percent) of their workday:** *Work Trend Index Annual Report*, Microsoft, May 9, 2023, https://www.microsoft.com/en-us/worklab/work-trend-index/will-ai-fix-work.

101 **One group of employees agreed to batch their email:** Indy Wijngaards, Florie R. Pronk, and Martijn J. Burger, "For Whom and Under What Circumstances Does Email Message Batching Work?" *Internet Interventions* 27 (2022): 100494.

102 **In a study of smartphone notification batching:** Nicholas Fitz et al., "Batching Smartphone Notifications Can Improve Well-Being," *Computers in Human Behavior* 101 (2019): 84–94.

102 **batching may actually increase:** Kathrin Reinke and Tomas Chamorro-Premuzic, "When Email Use Gets out of Control: Understanding the Relationship Between Personality and Email Overload and Their Impact on Burnout and Work Engagement," *Computers in Human Behavior* 36 (2014): 502–9.

104 **batching groups of activities:** Gloria Mark, Victor M. Gonzalez, and Justin

Harris, "No Task Left Behind? Examining the Nature of Fragmented Work," in *Proceedings of the SIGCHI Conference on Human Factors in Computing Systems* (Association for Computing Machinery, 2005), 321–30.

105 **creating a Sisyphean cycle:** Stephen R. Barley, Debra E. Meyerson, and Stine Grodal, "E-mail as a Source and Symbol of Stress," *Organization Science* 22, no. 4 (2011): 887–906.

106 **restricting email checking:** Laura A. Dabbish and Robert E. Kraut, "Email Overload at Work: An Analysis of Factors Associated with Email Strain," in *Proceedings of the 2006 20th Anniversary Conference on Computer Supported Cooperative Work* (Association for Computing Machinery, 2006), 431–40.

107 **they performed better when they streamed:** Bonnie Hayden Cheng, Yaxian Zhou, and Fangyuan Chen, "You've Got Mail! How Work E-mail Activity Helps Anxious Workers Enhance Performance Outcomes," *Journal of Vocational Behavior* 144 (2023): 103881.

RULE #4—WAIT. ONE HOUR. ONE DAY. ONE WEEK.

114 **what they call an "email urgency bias":** Laura M. Giurge and Vanessa K. Bohns, "You Don't Need to Answer Right Away! Receivers Overestimate How Quickly Senders Expect Responses to Non-Urgent Work Emails," *Organizational Behavior and Human Decision Processes* 167 (2021): 114–28.

115 **the more they report feeling email overload:** Andre Lanctot and Linda Duxbury, "Measurement of Perceived Importance and Urgency of Email: An Employees' Perspective," *Journal of Computer-Mediated Communication* 27, no. 2 (2022): zmac001.

115 **"Your Email Does Not Constitute My Emergency":** Adam Grant, "Your Email Does Not Constitute My Emergency," *New York Times*, April 13, 2023, https://www.nytimes.com/2023/04/13/opinion/email-time-work-stress.html.

118 **real patients triaged themselves too:** William C. Barley, Jeffrey W. Treem, and Paul M. Leonardi, "Experts at Coordination: Examining the Performance, Production, and Value of Process Expertise," *Journal of Communication* 70, no. 1 (2020): 60–89.

120 **"Being triaged might not feel much better":** Erica Dhawan, "Ignoring a Text Message or Email Isn't Always Rude. Sometimes It's Necessary," *New York Times,* February 21, 2022, https://www.nytimes.com/2022/02/21/opinion/culture/ghosting-work-digital-overload.html.

121 **we tend to entrain to other people's patterns:** Leyla Dogruel and Anna Schnauber-Stockmann, "What Determines Instant Messaging Communication? Examining the Impact of Person- and Situation-Level Factors on IM Responsiveness," *Mobile Media & Communication* 9, no. 2 (2021): 210–28; Joshua R. Tyler and John C. Tang, "When Can I Expect an Email Response? A Study of Rhythms in Email Usage," in *ECSCW 2003: Proceedings of the Eighth European Conference on Computer Supported Cooperative Work, 14–18 September 2003, Helsinki, Finland* (Springer, Netherlands, 2003), 239–58; Laura A. Dabbish et al., "Understanding Email Use: Predicting Action on a Message," in *Proceedings*

of the SIGCHI Conference on Human Factors in Computing Systems (Association for Computing Machinery, 2005), 691–700.

121 **even studies of preschoolers show:** Daniel J. Carroll, Emma Blakey, and Andrew Simpson, "Can We Boost Preschoolers' Inhibitory Performance Just by Changing the Way They Respond?," *Child Development* 92, no. 6 (2021): 2205–12.

123 **use digital tools effectively to maintain relationships:** Several of Walther's papers are cited in the seventh note in the Rule #2 chapter. His most concise message on this point can be found here: Joseph B. Walther, "Selective Self-Presentation in Computer-Mediated Communication: Hyperpersonal Dimensions of Technology, Language, and Cognition," *Computers in Human Behavior* 23, no. 5 (2007): 2538–57.

124 **eight hundred million emails:** Tarfah Alrashed, Ahmed Hassan Awadallah, and Susan Dumais, "The Lifetime of Email Messages: A Large-Scale Analysis of Email Revisitation," in *Proceedings of the 2018 Conference on Human Information Interaction & Retrieval* (Association for Computing Machinery, 2018), 120–29.

125 **Cal Newport professes a similar practice:** Cal Newport, "It's Okay to be Bad at E-mail," *Study Hacks Blog*, November 19, 2014, https://calnewport.com/its-okay-to-be-bad-at-e-mail.

127 **setting expectations about response time:** Ward van Zoonen, Anu Sivunen, and Jeffrey W. Treem, "Why People Engage in Supplemental Work: The Role of Technology, Response Expectations, and Communication Persistence," *Journal of Organizational Behavior* 42, no. 7 (2021): 867–84.

RULE #5—DON'T ASSUME

130 **The study of assumptions occupies a central role:** The magnum opus for finding the research on assumptions in human inference is: Lee Ross and Richard E. Nisbett, *The Person and the Situation: Perspectives of Social Psychology* (Pinter & Martin, 2011).

130 **ladder of inference:** The concept was developed by Harvard professor Chris Argyris and is put most to the point here: Chris Argyris, "The Executive Mind and Double-Loop Learning," *Organizational Dynamics* 11, no. 2 (1982): 5–22.

131 **he found a mess:** Paul M. Leonardi and Jeffrey W. Treem, "Knowledge Management Technology as a Stage for Strategic Self-Presentation: Implications for Knowledge Sharing in Organizations," *Information and Organization* 22, no. 1 (2012): 37–59.

134 **playground for cognitive biases:** Ofir Turel and Alexander Serenko, "Cognitive Biases and Excessive Use of Social Media: The Facebook Implicit Associations Test (FIAT)," *Addictive Behaviors* 105 (2020): 106328; Jihye Lee and James T. Hamilton, "Anchoring in the Past, Tweeting from the Present: Cognitive Bias in Journalists' Word Choices," *PLOS One* 17, no. 3 (2022): e0263730; Emily Dent and Andrew K. Martin, "Negative Comments and Social Media: How Cognitive Biases Relate to Body Image Concerns," *Body Image* 45 (2023): 54–64.

135 **increased feelings of well-being:** Li Sun, "Social Media Usage and Students' Social Anxiety, Loneliness and Well-Being: Does Digital Mindfulness-Based Intervention Effectively Work?," *BMC Psychology* 11, no. 1 (2023); Fengxia Lai et al., "Relationship Between Social Media Use and Social Anxiety in College Students: Mediation Effect of Communication Capacity," *International Journal of Environmental Research and Public Health* 20, no. 4 (2023): 3657.

135 **participants took a one-week break from Facebook:** Christopher G. Davis and Gary S. Goldfield, "Limiting Social Media Use Decreases Depression, Anxiety, and Fear of Missing Out in Youth with Emotional Distress: A Randomized Controlled Trial," *Psychology of Popular Media* (2024).

135 **asked to reduce their smartphone usage:** Davis and Goldfield, "Limiting Social Media Use Decreases Depression, Anxiety and Fear of Missing Out in Youth with Emotional Distress."

137 **called "theory of mind":** Rebecca Saxe, "How We Read Each Other's Minds," TED Talk, TED Global, July 2009, 16 min., 37 sec., https://www.ted.com/talks /rebecca_saxe_how_we_read_each_other_s_minds.

138 **Actively working to take the perspective:** Nic Hooper et al., "Perspective Taking Reduces the Fundamental Attribution Error," *Journal of Contextual Behavioral Science* 4, no. 2 (2015): 69–72; C. Daniel Batson, Shannon Early, and Giovanni Salvarani, "Perspective Taking: Imagining How Another Feels Versus Imaging How You Would Feel," *Personality and Social Psychology Bulletin* 23, no. 7 (1997): 751–58.

139 **an early AI agent named Lisa:** Camille G. Endacott and Paul M. Leonardi, "Artificial Intelligence and Impression Management: Consequences of Autonomous Conversational Agents Communicating on One's Behalf," *Human Communication Research* 48, no. 3 (2022): 462–90.

141 **Other studies have shown similar effects:** One study found that people blamed the individual they believed controlled a computer avatar, giving negative feedback more harshly than those who thought it was a program: Aike C. Horstmann, Jonathan Gratch, and Nicole C. Krämer, "I Just Wanna Blame Somebody, Not Something! Reactions to a Computer Agent Giving Negative Feedback Based on the Instructions of a Person," *International Journal of Human-Computer Studies* 154 (2021): 102683. Another study showed that receiving AI-enabled smart email responses improved perceptions of the sender's competence: Jess Hohenstein and Malte Jung, "AI as a Moral Crumple Zone: The Effects of AI-Mediated Communication on Attribution and Trust," *Computers in Human Behavior* 106 (2020): 106190.

RULE #6—ACT WITH INTENTION

145 **41 percent of internet users:** We Are Social and Meltwater, *Digital 2023 Global Overview Report*, DataReportal, January 26, 2023, https://datareportal .com/reports/digital-2023-global-overview-report.

145 **browsing social media and websites with little purpose:** Emily A. Vogels, Risa Gelles-Watnick, and Navid Massarat, *Teens, Social Media and Technology 2022*, Pew Research Center, August 10, 2022. In his book *The Anxious Genera-*

tion, Jonathan Haidt argues that this number is likely an underestimation and that teens are more likely on their devices roughly sixteen hours per day.

145 **roughly equivalent in many other countries:** For a great overview of lots of compiled stats on these topics, see Rob Binns, "Screen Time Statistics 2024," *Independent*, June 18, 2024, https://www.independent.co.uk/advisor/vpn/screen -time-statistics.

145 **In one of my favorite business books:** Teresa Amabile and Steven Kramer, *The Progress Principle: Using Small Wins to Ignite Joy, Engagement, and Creativity at Work* (Harvard Business, 2011).

146 **"job design theory":** J. Richard Hackman and Greg R. Oldham, "Development of the Job Diagnostic Survey," *Journal of Applied Psychology* 60, no. 2 (1975): 159; J. Richard Hackman and Greg R. Oldham, "Motivation through the Design of Work: Test of a Theory," *Organizational Behavior and Human Performance* 16, no. 2 (1976): 250–79.

148 **the concept of mindfulness:** Ellen J. Langer, *Mindfulness* (Da Capo, 2014).

148 **setting goals for their digital technology use:** Jason Bennett Thatcher et al., "Mindfulness in Information Technology Use," *MIS Quarterly* 42, no. 3 (2018): 831–48; Athina Ioannou, Mark Lycett, and Alaa Marshan, "The Role of Mindfulness in Mitigating the Negative Consequences of Technostress," *Information Systems Frontiers* 26, no. 2 (2024): 523–49; Elizabeth Marsh, Elvira Perez Vallejos, and Alexa Spence, "Mindfully and Confidently Digital: A Mixed Methods Study on Personal Resources to Mitigate the Dark Side of Digital Working," *PLOS One* 19, no. 2 (2024): e0295631.

150 **"Our willpower is limited":** Adam Alter, "Adam Alter: Irresistible Technology," interview by Alexandra Dempsey, Freedom, March 21, 2017, https://freedom .to/blog/adam-alter-irresistible.

150 **notifications drive our interactions with smartphones:** Maxi Heitmayer and Saadi Lahlou, "Why Are Smartphones Disruptive? An Empirical Study of Smartphone Use in Real-Life Contexts," *Computers in Human Behavior* 116 (2021): 106637.

151 **texting with people during the surgery:** Jamie E. Guillory et al., "Text Messaging Reduces Analgesic Requirements During Surgery," *Pain Medicine* 16, no. 4 (2015): 667–72.

151 **Chronic smartphone use changes our brain chemistry:** Radiological Society of North America, "Smartphone Addiction Creates Imbalance in Brain, Study Suggests," ScienceDaily, November 30, 2017, https://www.sciencedaily .com/releases/2017/11/171130090041.htm.

151 **Drugs like heroin cause an increase in GABA:** Zheng-Xiong Xi et al., "GABAergic Mechanisms of Heroin-Induced Brain Activation Assessed with Functional MRI," *Magnetic Resonance in Medicine: An Official Journal of the International Society for Magnetic Resonance in Medicine* 48, no. 5 (2002): 838–43.

151 **"the modern-day hypodermic needle":** Anna Lembke, *Dopamine Nation: Finding Balance in the Age of Indulgence* (Penguin, 2021), 1.

151 **Habits are largely the result of unconscious action:** Charles Duhigg, *The Power of Habit: Why We Do What We Do in Life and Business* (Random House, 2012); James Clear, *Atomic Habits: An Easy & Proven Way to Build Good Habits*

& *Break Bad Ones* (Avery, 2018). For a more scientific view, see Ann M. Graybiel, "Habits, Rituals, and the Evaluative Brain," *Annual Review of Neuroscience* 31, no. 1 (2008): 359–87.

152 **habits that people have a hard time breaking:** Antti Oulasvirta et al., "Habits Make Smartphone Use More Pervasive," *Personal and Ubiquitous Computing* 16 (2012): 105–14; Arun Vishwanath, "Habitual Facebook Use and Its Impact on Getting Deceived on Social Media," *Journal of Computer-Mediated Communication* 20, no. 1 (2015): 83–98; Jean-Charles Pillet and Kevin Daniel André Carillo, "Email-Free Collaboration: An Exploratory Study on the Formation of New Work Habits Among Knowledge Workers," *International Journal of Information Management* 36, no. 1 (2016): 113–25.

153 **from habits of mind to active thinking:** Meryl Reis Louis and Robert I. Sutton, "Switching Cognitive Gears: From Habits of Mind to Active Thinking," *Human Relations* 44, no. 1 (1991): 55–76.

156 **offering phone-free trips:** Christine Chung, "A Girl's Trip to Costa Rica but with No Phones. Did It Happen?," *New York Times,* June 5, 2024, https://www.nytimes.com/2024/06/05/travel/phone-free-tours-costa-rica-girls-trip.html.

RULE #7—LEARN VICARIOUSLY

160 **improved their ability to find experts:** Paul M. Leonardi, "Social Media, Knowledge Sharing, and Innovation: Toward a Theory of Communication Visibility," *Information Systems Research* 25, no. 4 (2014): 796–816; Paul M. Leonardi, "Ambient Awareness and Knowledge Acquisition," *MIS Quarterly* 39, no. 4 (2015): 747–62.

161 **long-standing line of research I've conducted:** Paul M. Leonardi and Jeffrey W. Treem, "Behavioral Visibility: A New Paradigm for Organization Studies in the Age of Digitization, Digitalization, and Datafication," *Organization Studies* 41, no. 12 (2020): 1601–25.

161 **Bonnie Nardi and Yrjö Engeström:** Bonnie A. Nardi and Yrjö Engeström, "A Web on the Wind: The Structure of Invisible Work," *Computer Supported Cooperative Work: The Journal of Collaborative Computing* 8 (1999): 1–8.

163 **people often start their work as apprentices:** Jean Lave and Etienne Wenger, *Situated Learning: Legitimate Peripheral Participation* (Cambridge University Press, 1991).

163 **accurate metaknowledge is associated with:** Yuqing Ren, Kathleen M. Carley, and Linda Argote, "The Contingent Effects of Transactive Memory: When Is It More Beneficial to Know What Others Know?," *Management Science* 52, no. 5 (2006): 671–82.

166 **150 meaningful connections:** Robin Dunbar, "Neocortex Size as a Constraint on Group Size in Primates," *Journal of Human Evolution* 22, no. 6 (1992): 469–93.

166 **"majority of what people see on the platform":** Brendan Nyhan et al., "Like-Minded Sources on Facebook Are Prevalent but Not Polarizing," *Nature* 620, no. 7972 (2023): 137–44.

167 **number and diversity of connections declined:** Longqi Yang et al., "The Effects of Remote Work on Collaboration among Information Workers," *Nature Human Behaviour* 6, no. 1 (2022): 43–54.

167 **possibilities for us to learn vicariously:** Luke Rhee and Paul M. Leonardi, "Which Pathway to Good Ideas? An Attention-Based View of Innovation in Social Networks," *Strategic Management Journal* 39, no. 4 (2018): 1188–1215; Luke Rhee and Paul Leonardi, "Borrowing Networks for Innovation: The Role of Attention Allocation in Secondhand Brokerage," *Strategic Management Journal* 45, no. 1 (2024): 1326–65.

168 **combining ideas in new ways:** Andrew Hargadon, *How Breakthroughs Happen: The Surprising Truth About How Companies Innovate* (Harvard Business School Press, 2003).

169 **study of the video game *World of Warcraft*:** Bonnie Nardi, *My Life as a Night Elf Priest: An Anthropological Account of World of Warcraft* (University of Michigan Press, 2010).

170 **"each little update":** Clive Thompson, "Brave New World of Digital Intimacy," *New York Times Magazine*, September 5, 2008, https://www.nytimes.com/2008/09/07/magazine/07awareness-t.html.

RULE #8—BE HERE, NOT ELSEWHERE

176 **a concept called "flow":** Mihaly Csikszentmihalyi, *Flow: The Psychology of Optimal Experience* (Harper Perennial, 2008).

176 **sense of deep enjoyment and fulfillment:** Thais Piassa Rogatko, "The Influence of Flow on Positive Affect in College Students," *Journal of Happiness Studies* 10 (2009): 133–48; A. L. Collins, Natalia Sarkisian, and Ellen Winner, "Flow and Happiness in Later Life: An Investigation into the Role of Daily and Weekly Flow Experiences," *Journal of Happiness Studies* 10 (2009): 703–19; Hsiang Chen, "Flow on the Net: Detecting Web Users' Positive Affects and Their Flow States," *Computers in Human Behavior* 22, no. 2 (2006): 221–33.

178 **described by the Yerkes-Dodson law:** Robert Mearns Yerkes and John D. Dodson, "The Relation of Strength of Stimulus to Rapidity of Habit-Formation," *Journal of Comparative Neurology and Psychology* 18, no. 5 (1908): 459–82.

179 **arousal for peak performance:** For more discussion about this, see Brad Stulberg and Steve Magness, *Peak Performance: Elevate Your Game, Avoid Burnout, and Thrive with the New Science of Success* (Rodale, 2017); Amishi Jha, *Peak Mind: Find Your Focus, Own Your Attention, Invest 12 Minutes a Day* (Hachette, 2021).

180 **buildup of something called glutamate:** Mia Michaela Pal, "Glutamate: The Master Neurotransmitter and Its Implications in Chronic Stress and Mood Disorders," *Frontiers in Human Neuroscience* 15 (2021): 722323.

180 **the possibility of being "forever elsewhere":** Sherry Turkle, *Reclaiming Conversation: The Power of Talk in a Digital Age* (Penguin, 2016), 3.

181 **flow when using their digital tools:** Ritu Agarwal and Elena Karahanna, "Time Flies When You're Having Fun: Cognitive Absorption and Beliefs About Information Technology Usage," *MIS Quarterly* (2000): 665–94; Jane Webster,

Linda Klebe Trevino, and Lisa Ryan, "The Dimensionality and Correlates of Flow in Human-Computer Interactions," *Computers in Human Behavior* 9, no. 4 (1993): 411–26; Linda Klebe Trevino and Jane Webster, "Flow in Computer-Mediated Communication: Electronic Mail and Voice Mail Evaluation and Impacts," *Communication Research* 19, no. 5 (1992): 539–73.

182 **biggest effects are found among video game players:** Alasdair G. Thin, Lisa Hansen, and Danny McEachen, "Flow Experience and Mood States While Playing Body Movement-Controlled Video Games," *Games and Culture* 6, no. 5 (2011): 414–28; Alistair Raymond Bryce Soutter and Michael Hitchens, "The Relationship Between Character Identification and Flow State Within Video Games," *Computers in Human Behavior* 55 (2016): 1030–38; Seung-A Annie Jin, "'I Feel Present. Therefore, I Experience Flow:' A Structural Equation Modeling Approach to Flow and Presence in Video Games," *Journal of Broadcasting & Electronic Media* 55, no. 1 (2011): 114–36.

182 **people who use social media mindfully:** Julia Brailovskaia and Jürgen Margraf, "From Fear of Missing Out (FoMO) to Addictive Social Media Use: The Role of Social Media Flow and Mindfulness," *Computers in Human Behavior* 150 (2024): 107984.

185 **"increased vigor and reduced emotional exhaustion":** Hongjai Rhee and Sudong Kim, "Effects of Breaks on Regaining Vitality at Work: An Empirical Comparison of 'Conventional' and 'Smart Phone' Breaks," *Computers in Human Behavior* 57 (2016): 160–67.

185 **Research shows that spending time outside:** Anthony C. Klotz et al., "Getting Outdoors After the Workday: The Affective and Cognitive Effects of Evening Nature Contact," *Journal of Management* 49, no. 7 (2023): 2254–87.

185 **Spending 1 minute and 40 seconds:** Richard G. Coss and Craig M. Keller, "Transient Decreases in Blood Pressure and Heart Rate with Increased Subjective Level of Relaxation While Viewing Water Compared with Adjacent Ground," *Journal of Environmental Psychology* 81 (2022): 101794.

186 **In a result that will surprise probably no one:** Although the effects of engagement didn't quite reach the required level in significance testing to hang your hat on (perhaps, as the authors suggest, because of the difficulty and awkwardness of reporting on the quality of our sexual experiences), they are still suggestive. The authors did find that experiencing high levels of stress and exhaustion in a given workday made it less likely that a person would have sex that evening that could help them reduce their exhaustion during the next day. Keith Leavitt et al., "From the Bedroom to the Office: Workplace Spillover Effects of Sexual Activity at Home," *Journal of Management* 45, no. 3 (2019): 1173–92.

187 **makes my digital tools seem unattractive to me:** I like the advice of this article by Melissa Kirsch, who suggests one place to look for complementary opposites is your past childhood hobbies: Melissa Kirsch, "Old Skills, New Rewards," *New York Times*, March 4, 2023, https://www.nytimes.com/2023/03/04/briefing/trying-activities-again.html.

189 **Disconnecting doesn't lead directly to rejuvenation:** Shawn Achor and Michelle Gielan, "Resilience Is About How You Recharge, Not How You Endure," *Harvard Business Review,* June 24, 2016, https://hbr.org/2016/06/resilience

-is-about-how-you-recharge-not-how-you-endure. See also Shawn Achor, *The Happiness Advantage: How a Positive Brain Fuels Success in Work and Life* (Crown Currency, 2010).

189 **people who set goals for their days off:** Laura M. Giurge and Vanessa Bohns, "Be Intentional About How You Spend Your Time Off," *Harvard Business Review,* December 1, 2021, https://hbr.org/2021/12/be-intentional-about-how-you -spend-your-time-off.

189 **reported higher levels of exhaustion and lower levels of energy**: Ciara M. Kelly et al., "The Relationship Between Leisure Activities and Psychological Resources That Support a Sustainable Career: The Role of Leisure Seriousness and Work-Leisure Similarity," *Journal of Vocational Behavior* 117 (2020): 103340.

190 **findings of thirteen separate studies:** Gabriela N. Tonietto and Selin A. Malkoc, "The Calendar Mindset: Scheduling Takes the Fun Out and Puts the Work In," *Journal of Marketing Research* 53, no. 6 (2016): 922–36.

CHAPTER 4. HOW NOT TO BE AN ENERGY VAMPIRE: LESSONS FOR WORKPLACE MANAGERS

194 **employees don't give a lot of credence:** Dorothy Leonard-Barton and Isabelle Deschamps, "Managerial Influence in the Implementation of New Technology," *Management Science* 34, no. 10 (1988): 1252–65.

195 **Microsoft's Copilot (which is powered by OpenAI:** See some of the analysis of how Microsoft's Copilot is changing work in Microsoft's 2023 Work Index Report, August 2, 2023, Microsoft News Center, https://news.microsoft. com/en-xm/2023/08/02/microsofts-2023-work-trend-index-report-reveals-impact-of-digital-debt-on-innovation-emphasizes-need-for-ai-proficiency-for-every-employee.

195 **"I don't want to go back to life without it":** Jared Spataro, "Does your email inbox ever resemble a mountain you just can't summit?," LinkedIn, October 23, 2023, https://www.linkedin.com/posts/jaredspa_worktrendindex-microsoft365 copilot-generativeai-activity-7122239600076472322-uyq2.

197 **"land and expand" sales strategy:** Thanks to Kenny Van Zant, cofounder of Asana and advisor to many SaaS companies, for discussing pricing strategy with me.

200 **early research projects was with General Motors:** For an in-depth overview of this work, see Paul M. Leonardi, *Car Crashes without Cars: Lessons about Simulation Technology and Organizational Change from Automotive Design* (MIT Press, 2012).

201 **Here are two examples:** More context for these examples can be found in: Paul M. Leonardi, "When Does Technology Use Enable Network Change in Organizations? A Comparative Study of Feature Use and Shared Affordances," *MIS Quarterly* 37, no. 3 (2013): 749–75.

203 **the power of managerial framing around new technologies:** Wanda J. Orlikowski and Debra C. Gash, "Technological Frames: Making Sense of Information Technology in Organizations," *ACM Transactions on Information*

Systems (TOIS) 12, no. 2 (1994): 174–207; Elizabeth Davidson, "A Technological Frames Perspective on Information Technology and Organizational Change," *The Journal of Applied Behavioral Science* 42, no. 1 (2006): 23–39; Amy C. Edmondson, Richard M. Bohmer, and Gary P. Pisano, "Disrupted Routines: Team Learning and New Technology Implementation in Hospitals," *Administrative Science Quarterly* 46, no. 4 (2001): 685–716.

203 **"However, something usually happens":** Stephen R. Barley, *Work and Technological Change* (Oxford University Press, 2020), 26.

207 **they meet face-to-face in the office periodically:** Pamela J. Hinds and Catherine Durnell Cramton, "Situated Coworker Familiarity: How Site Visits Transform Relationships Among Distributed Workers," *Organization Science* 25, no. 3 (2014): 794–814.

207 ***Remote Work Revolution:*** Tsedal Neeley, *Remote Work Revolution: Succeeding from Anywhere* (Harper Business, 2021).

208 **report lower levels of trust:** Prasert Kanawattanachai and Youngjin Yoo, "Dynamic Nature of Trust in Virtual Teams," *Journal of Strategic Information Systems* 11, no. 3–4 (2002): 187–213; Christina Breuer, Joachim Hüffmeier, and Guido Hertel, "Does Trust Matter More in Virtual Teams? A Meta-analysis of Trust and Team Effectiveness Considering Virtuality and Documentation as Moderators," *Journal of Applied Psychology* 101, no. 8 (2016): 1151–77.

208 **good managers can boost feelings of camaraderie:** Tsedal B. Neeley and Paul M. Leonardi, "Enacting Knowledge Strategy Through Social Media: Passable Trust and the Paradox of Nonwork Interactions," *Strategic Management Journal* 39, no. 3 (2018): 922–46.

211 **This framework, called STEP:** The following material on the STEP framework is adapted and reprinted with permission from Paul M. Leonardi, "Helping Employees Succeed with Generative AI: How to Manage Performance When New Technology Upends Traditional Business Processes," *Harvard Business Review* 10, no. 6 (2023): 49–53.

212 **Estimates suggest up to 80 percent:** Tyna Eloundou et al., "GPTs Are GPTs: An Early Look at the Labor Market Impact Potential of Large Language Models," preprint, arXiv, March 17, 2023, arXiv:2303.10130.

212 **determine which tasks AI would not be helpful for:** They did this by adopting a learning frame in their talk about AI, as suggested by Amy Edmondson: Amy C. Edmondson, "Framing for Learning: Lessons in Successful Technology Implementation," *California Management Review* 45, no. 2 (2003): 34–54.

215 **what is often called "fine-tuning":** Fine-tuning is a machine learning technique used to refine a pretrained model by training it further on a smaller, specialized dataset. This approach allows the model to adapt its weights and biases to the nuances of the new data without starting from scratch. Prompt engineering involves crafting and optimizing the inputs (prompts) given to a machine learning model, typically a large language model, to guide it in generating specific desired outputs. It's akin to framing a question or providing context in a way that makes it more likely to receive the right answer.

215 **need for continuous employee reskilling:** If you are interested in how you

might think about setting up a reskilling program to help employees develop a digital mindset, I provide examples in Paul M. Leonardi and Tsedal Neeley, *The Digital Mindset: What It Really Takes to Thrive in the Age of Data, Algorithms, and AI* (Harvard Business Review Press, 2022).

217 **are likely to cascade:** For a detailed account of how and why technology-induced role changes are likely to shift patterns of interactions across an organization, see Stephen R. Barley, "The Alignment of Technology and Structure Through Roles and Networks," *Administrative Science Quarterly* 35 (1990): 61–103.

CHAPTER 5. PUT ON YOUR OWN MASK BEFORE HELPING OTHERS: A GUIDE FOR PARENTS

221 **look to the notes:** For books on technology and adolescence, I recommend: Jean M. Twenge, *iGen: Why Today's Super-Connected Kids Are Growing Up Less Rebellious, More Tolerant, Less Happy—and Completely Unprepared for Adulthood—and What That Means for the Rest of Us* (Simon & Schuster, 2017); Jonathan Haidt, *The Anxious Generation: How the Great Rewiring of Childhood Is Causing an Epidemic of Mental Illness* (Random House, 2024).

224 **A robust line of research shows:** Correlated: Aurélie Gillis and Isabelle Roskam, "Daily Exhaustion and Support in Parenting: Impact on the Quality of the Parent–Child Relationship," *Journal of Child and Family Studies* 28 (2019): 2007–16. Causal: Moïra Mikolajczak, James J. Gross, and Isabelle Roskam, "Parental Burnout: What Is It, and Why Does It Matter?," *Clinical Psychological Science* 7, no. 6 (2019): 1319–29. For a detailed review of studies showing both correlation and causation, see Moïra Mikolajczak and Isabelle Roskam, "Parental Burnout: Moving the Focus from Children to Parents," *New Directions for Child and Adolescent Development* 2020, no. 174 (2020): 7–13.

225 **there is a Parkinson's law:** Parkinson's law is the old adage that work expands to fill the time allotted for its completion. The term was first coined by Cyril Northcote Parkinson in a humorous essay he wrote for *The Economist* in 1955. He shares the story of a woman whose only task in a day is to send a postcard—a task which would take a busy person approximately three minutes. But the woman spends an hour finding the card, another half hour looking for her glasses, ninety minutes writing the card, twenty minutes deciding whether or not to take an umbrella along on her walk to the mailbox . . . and on and on until her day is filled.

226 **that means only half an hour was saved:** Paul M. Leonardi and Diane E. Bailey, "Transformational Technologies and the Creation of New Work Practices: Making Implicit Knowledge Explicit in Task-Based Offshoring," *MIS Quarterly* 32, no. 2 (2008): 411–36.

227 **But they never went away completely:** As my coauthors and I explain in a more detailed analysis, all the specification in the world couldn't eliminate the difficulty of trying to determine whether a simulation model was accurate when the physical parts of the vehicle in question were thousands of miles away:

Diane E. Bailey, Paul M. Leonardi, and Stephen R. Barley, "The Lure of the Virtual," *Organization Science* 23, no. 5 (2012): 1485–1504.

230 **significant driver of "parental burnout":** Moïra Mikolajczak et al., "Exhausted Parents: Sociodemographic, Child-Related, Parent-Related, Parenting, and Family-Functioning Correlates of Parental Burnout," *Journal of Child and Family Studies* 27 (2018): 602–14.

230 **In the immediate wake of this outage:** Tal Eitan and Tali Gazit, "No Social Media for Six Hours? The Emotional Experience of Meta's Global Outage According to FoMO, JoMO, and Internet Intensity," *Computers in Human Behavior* 138 (2023): 107474.

231 **trade FOMO for JOMO:** Tal Eitan and Tali Gazit, "The 'Here and Now' Effect: JoMO, FoMO, and the Well-Being of Social Media Users," *Online Information Review* (2024).

231 **how people can experience JOMO:** Steven S. Chan et al., "Social Media and Mindfulness: From the Fear of Missing Out (FOMO) to the Joy of Missing Out (JOMO)," *Journal of Consumer Affairs* 56, no. 3 (2022): 1312–31.

232 **turned off the push notifications:** This study, which we looked at in Rule #3 on batching, shows this is a good tactic for reducing FOMO: Nicholas Fitz et al., "Batching Smartphone Notifications Can Improve Well-Being," *Computers in Human Behavior* 101 (2019): 84–94.

233 **a popular way to measure tie strength:** Morten T. Hansen, "The Search-Transfer Problem: The Role of Weak Ties in Sharing Knowledge Across Organization Subunits," *Administrative Science Quarterly* 44, no. 1 (1999): 82–111.

235 **In a recent study led by Jeffrey Hall:** Jeffrey A. Hall et al., "Social Bandwidth: When and Why Are Social Interactions Energy Intensive?," *Journal of Social and Personal Relationships* 40, no. 8 (2023): 2614–36.

237 **they tend to pay less attention to their kids:** For two comprehensive reviews of this topic, see Brandon T. McDaniel, "Parent Distraction with Phones, Reasons for Use, and Impacts on Parenting and Child Outcomes: A Review of the Emerging Research," *Human Behavior and Emerging Technologies* 1, no. 2 (2019): 72–80; Cory A. Kildare and Wendy Middlemiss, "Impact of Parents' Mobile Device Use on Parent-Child Interaction: A Literature Review," *Computers in Human Behavior* 75 (2017): 579–93.

237 **children tend to either withdraw or act out:** Xingchao Wang et al., "Parental Phubbing and Children's Social Withdrawal and Aggression: A Moderated Mediation Model of Parenting Behaviors and Parents' Gender," *Journal of Interpersonal Violence* 37, no. 21–22 (2022): 19395–419.

237 **the worse their kids behaved:** Brandon T. McDaniel and Jenny S. Radesky, "Technoference: Longitudinal Associations Between Parent Technology Use, Parenting Stress, and Child Behavior Problems," *Pediatric Research* 84, no. 2 (2018): 210–18.

237 **increased technology interference in parenting:** Genni Newsham, Michelle Drouin, and Brandon T. McDaniel, "Problematic Phone Use, Depression, and Technology Interference Among Mothers," *Psychology of Popular Media* 9, no. 2 (2020): 117–24.

237 **a systematic review of twenty-seven studies:** Cory A. Kildare and Wendy Middlemiss, "Impact of Parents Mobile Device Use on Parent-Child Interaction: A Literature Review," *Computers in Human Behavior* 75 (2017): 579–93.

238 **"a message to their children about what's important":** Quoted in Keith Hamm, "Study Finds Parents' Phone Use in Front of Their Kids Can Harm Emotional Intelligence," *Current*, March 10, 2023, news.ucsb.edu/2023/020867 /screen-time-concerns.

239 **being on a device all the time was OK:** Lindsay Blackwell, Emma Gardiner, and Sarita Schoenebeck, "Managing Expectations: Technology Tensions among Parents and Teens," in *Proceedings of the 19th ACM Conference on Computer-Supported Cooperative Work & Social Computing* (Association for Computing Machinery, 2016), 1390–401. This study also found that parents thought that their use of the technology led to their kids having increased screen time and they felt guilty for that: Lara N. Wolfers, Robin L. Nabi, and Nathan Walter, "Too Much Screen Time or Too Much Guilt? How Child Screen Time and Parental Screen Guilt Affect Parental Stress and Relationship Satisfaction," *Media Psychology* (2024): 1–32.

CHAPTER 6. THE ARTIFICE OF INTELLIGENCE: LIVING AND WORKING WITH AI

243 **Noam Chomsky, famed MIT linguistics professor:** Noam Chomsky, Ian Roberts, and Jeffrey Watumull, "Noam Chomsky: The False Promise of Chat-GPT," *New York Times*, March 8, 2023, https://www.nytimes.com/2023/03/08/ opinion/noam-chomsky-chatgpt-ai.html.

244 **the witty words of French sociologist Jean Baudrillard:** Jean Baudrillard, *The Transparency of Evil: Essays on Extreme Phenomena* (Verso Books, 2000), 58.

244 **formidable prediction machines:** For an in-depth explanation of why it is useful to think of AI as a prediction machine, see Ajay Agrawal, Joshua Gans, and Avi Goldfarb, *Prediction Machines, Updated and Expanded: The Simple Economics of Artificial Intelligence* (Harvard Business Review Press, 2022).

244 **"spontaneously emerged" as the algorithms:** Michal Kosinski, "Evaluating Large Language Models in Theory of Mind Tasks," *Computer Sciences* 121, no. 45 (October 29, 2024), https://www.pnas.org/doi/10.1073/pnas.2405460121.

246 **One field experiment:** Fabrizio Dell'Acqua et al., "Navigating the Jagged Technological Frontier: Field Experimental Evidence of the Effects of AI on Knowledge Worker Productivity and Quality," Working Paper No. 24-013 (Technology & Operations Management Unit, Harvard Business School, September 2023).

246 **Another survey by Microsoft:** "What Can Copilot's Earliest Users Teach Us About Generative AI at Work?," *Work Trend Index Special Report, Microsoft,* November 15, 2023, https://www.microsoft.com/en-us/worklab/work-trend-index /copilots-earliest-users-teach-us-about-generative-ai-at-work.

246 **Another study conducted by researchers at OpenAI:** Tyna Eloundou et al., "GPTs Are GPTs: An Early Look at the Labor Market Impact Potential of Large Language Models," preprint, arXiv, March 17, 2023, arXiv:2303.10130.

247 **the vast majority of content on the web will be produced by AI:** Europol Innovation Lab, *Facing Reality? Law Enforcement and the Challenge of Deepfakes* (Publications Office of the European Union, 2022), https://www.europol.europa .eu/publications-events/publications/facing-reality-law-enforcement-and -challenge-of-deepfakes.

250 **Asana and Anthropic revealed:** Rebecca Hinds et al., *The State of AI at Work* (Asana Work Innovation Lab and Anthropic, 2024), https://asana.com/work -innovation-lab/state-of-ai-at-work.

252 **That same Microsoft study that suggested:** "What Can Copilot's Earliest Users Teach Us About Generative AI at Work?"

253 **brilliant piece of investigative reporting:** Kevin Schaul, Szu Yu Chen, and Nitasha Tiku, "Inside the Secret List of Websites That Make AI like ChatGPT Sound Smart," *Washington Post,* April 19, 2023, https://www.washingtonpost .com/technology/interactive/2023/ai-chatbot-learning.

255 **these innovators have a "vision advantage":** Ronald S. Burt, "Structural Holes and Good Ideas," *American Journal of Sociology* 110, no. 2 (2004): 349–99.

255 **end up spanning social groups for so many reasons:** Prasad Balkundi et al., "Demographic Antecedents and Performance Consequences of Structural Holes in Work Teams," *Journal of Organizational Behavior: The International Journal of Industrial, Occupational and Organizational Psychology and Behavior* 28, no. 2 (2007): 241–60; Akbar Zaheer and Giuseppe Soda, "The Evolution of Network Structure: Where Do Structural Holes Come From?," *Administrative Science Quarterly* 54, no. 1 (2009): 1–31; Ronald S. Burt, "Network-Related Personality and the Agency Question: Multirole Evidence from a Virtual World," *American Journal of Sociology* 118, no. 3 (2012): 543–91.

256 **connections necessary for innovation is often short-lived:** Paul M. Leonardi and Diane E. Bailey, "Recognizing and Selling Good Ideas: Network Articulation and the Making of an Offshore Innovation Hub," *Academy of Management Discoveries* 3, no. 2 (2017): 116–44; Eric Quintane et al., "Why Employees Who Work Across Silos Get Burned Out," *Harvard Business Review*, May 13, 2024, https://www.hbr.org/2024/05/why-employees-who-work-across-silos-get -burned-out.

256 **"They are trained by generating relationships between tokens":** Ethan Mollick, *Co-Intelligence: Living and Working with AI* (Penguin, 2024), 100.

256 **"predicts outsized impact":** Feng Shi and James Evans, "Surprising Combinations of Research Contents and Contexts Are Related to Impact and Emerge with Scientific Outsiders from Distant Disciplines," *Nature Communications* 14, no. 1 (2023): 1641–55.

257 **improves the work of poor performers:** Dell'Acqua et al., "Navigating the Jagged Technological Frontier."

257 **known as "model collapse":** See, for example, Ilia Shumailov et al., "The Curse of Recursion: Training on Generated Data Makes Models Forget," preprint, arXiv, May 27, 2023, arXiv:2305.17493; Matthias Gerstgrasser et al., "Is Model Collapse Inevitable? Breaking the Curse of Recursion by Accumulating Real and Synthetic Data," preprint, arXiv, April 1, 2024, arXiv:2404.01413.

257 **inevitable even under the best learning conditions:** For a great explanation of why model collapse is such a problem, see Ben Lutkevich, "Model Collapse Explained: How Synthetic Training Data Breaks AI," *Informa TechTarget*, July 7, 2023, https://www.techtarget.com/whatis/feature/Model-collapse-explained -How-synthetic-training-data-breaks-AI.

258 **AlphaFold, a system powered:** To learn more about AlphaFold, see "Artificial Intelligence Is Taking Over Drug Development," *Economist*, March 27, 2024, https://www.economist.com/technology-quarterly/2024/03/27/artificial -intelligence-is-taking-over-drug-development.

258 **a major milestone for AI-enabled drug discovery:** Steve Nouri, "Generative AI Drugs Are Coming," *Forbes*, September 5, 2023, https://www.forbes .com/sites/forbestechcouncil/2023/09/05/generative-ai-drugs-are-coming.

258 **helped researchers to discover 2.2 million new crystals:** Amil Merchant and Ekin Dogus Cubuk, "Millions of New Materials Discovered with Deep Learning," Google DeepMind, November 29, 2023, https://deepmind.google /discover/blog/millions-of-new-materials-discovered-with-deep-learning.

258 **seven hundred of these new structures in the lab:** Amil Merchant et al., "Scaling Deep Learning for Materials Discovery," *Nature* 624, no. 7990 (2023): 80–85.

259 **successfully killed a resistant strain of the bacteria:** Kyle Swanson et al., "Generative AI for Designing and Validating Easily Synthesizable and Structurally Novel Antibiotics," *Nature Machine Intelligence* 6, no. 3 (2024): 338–53; Radhika Rajkumar, "How AI Hallucinations Could Help Create Life-Saving Antibiotics," *ZDNET*, April 24, 2024, https://www.zdnet.com/article/how-ai -hallucinations-could-help-create-life-saving-antibiotics.

259 **beneficial AI hallucinations:** Ethan Mollick, *Co-Intelligence: Living and Working with AI* (Penguin, 2024).

260 **taking the idea of harnessing AI hallucinations:** Rebecca Bellen, "How Maven's AI-run 'Serendipity Network' Can Make Social Media Interesting Again," *Tech Crunch,* May 26, 2024, https://techcrunch.com/2024/05/26/how -mavens-ai-run-serendipity-network-can-make-social-media-interesting -again.

262 **restructure their workflows:** Thomas H. Davenport and Steven M. Miller, *Working with AI: Real Stories of Human-Machine Collaboration* (MIT Press, 2022).

262 **found what they were looking for faster:** "What Can Copilot's Earliest Users Teach Us About Generative AI at Work?"

263 **flow in activities that were not of our own choosing:** Roger C. Mannell, Jiri Zuzanek, and Reed Larson, "Leisure States and 'Flow' Experiences: Testing Perceived Freedom and Intrinsic Motivation Hypotheses," *Journal of Leisure Research* 20, no. 4 (1988): 289–304.

263 **proved to be surprisingly persuasive to the target:** Sandra C. Matz et al., "The Potential of Generative AI for Personalized Persuasion at Scale," *Scientific Reports* 14, no. 1 (2024): 4692.

263 **AI model exceeded human experts in accuracy:** Shotaro Doki et al., "Com-

parison of Predicted Psychological Distress Among Workers Between Artificial Intelligence and Psychiatrists: A Cross-Sectional Study in Tsukuba Science City, Japan," *BMJ Open* 11, no. 6 (2021): e046265.

264 **AI as a better communicator on nearly every dimension:** Tao Tu et al., "Towards Conversational Diagnostic AI," preprint, arXiv, January 11, 2024, arXiv:2401.05654.

INDEX

Page numbers in italics indicate images.